Data Processing with Optimus

Supercharge big data preparation tasks for analytics and machine learning with Optimus using Dask and PySpark

Dr. Argenis Leon

Luis Aguirre

BIRMINGHAM—MUMBAI

Data Processing with Optimus

Publishing Product Manager: Devika Battike
Senior Editor: David Sugarman
Content Development Editor: Joseph Sunil
Technical Editor: Manikandan Kurup
Copy Editor: Safis Editing
Project Coordinator: Aparna Ravikumar Nair
Proofreader: Safis Editing
Indexer: Pratik Shirodkar
Production Designer: Sinhayna Bais

First published: September 2021

Production reference: 1290721

Published by Packt Publishing Ltd.
Livery Place
35 Livery Street
Birmingham
B3 2PB, UK.

ISBN 978-1-80107-956-3

www.packt.com

Contributors

About the authors

Dr. Argenis Leon created Optimus, an open-source python library built over PySpark aimed to provide an easy-to-use API to clean, process, and merge data at scale. Since 2012, Argenis has been working on big data-related projects using Postgres, MongoDB, Elasticsearch for social media data collection and analytics. In 2015, he started working on Machine learning projects in Retail, AdTech, and Real Estate in Venezuela and Mexico. In 2019 he created Bumblebee, a low-code open-source web platform to clean and wrangle big data using CPU and GPUs using NVIDIA RAPIDS. Nowadays Argenis is Co-founder and CTO of boitas.com (backed by YCombinator) a wholesale marketplace for SMB in Latin America.

Luis Aguirre began working with web development projects for Mood Agency in 2018, creating sites for brands from all across Latin America. One year later he started working on Bumblebee, a low-code web platform to transform data that uses Optimus. In mid-2020 he started participating in the Optimus project as a core developer; focusing on creating the easiest-to-use experience for both projects. In 2021 he started working on the Optimus REST API, a tool to allow requests from the web focused on data wrangling.

About the reviewer

Sergio Sánchez Zavala, originally from Tijuana, Baja California, Mexico, is a self-described hip-hop head, public policy wonk, and data nerd. He is dedicated to making research, open source tools, and resources transparent, reproducible, and accessible. He is also the creator of @tacosdedatos, an online community for learning data analysis, engineering, and visualization best practices and techniques in Spanish. He is currently a data engineer at Alluma, a social enterprise in the social tech space providing people-centered, policy-driven technology solutions and consulting services to state, county, and local government agencies, nonprofits, and other partners.

Table of Contents

Section 2: Optimus – Transform and Rollout

3

Data Wrangling

4

Combining, Reshaping, and Aggregating Data

5
Data Visualization and Profiling

6
String Clustering

7
Feature Engineering

Section 3: Advanced Features of Optimus

8
Machine Learning

9
Natural Language Processing

10
Hacking Optimus

11

Optimus as a Web Service

Other Books You May Enjoy

Index

Preface

Optimus is a Python library that works as a unified API for data cleaning, processing, and merging. It can be used for small and big data on local and big clusters using CPUs or GPUs. *Data Processing with Optimus* shows you how to use the library to enhance your data science workflow.

The book begins by covering the internals of Optimus and showing you how it works in tandem with existing technologies to serve users' data processing needs. You'll then use Optimus to load and save data from text data formats such as CSV and JSON files, explore binary files such as Excel, and for columnar data processing with Parquet, Avro, and OCR. Next, you'll learn about the profiler and profiler data types, a unique feature of Optimus DataFrames that helps you get an overview of the data quality in every column. You'll also create data cleaning and transformation functions and add a hypothetical new data processing engine. Later, you'll explore plots in Optimus such as histograms and box plots, and learn how Optimus lets you connect to any other library, including Plotly and Altair. Finally, you'll understand the advanced applications of Optimus, such as feature engineering, machine learning, and NLP, along with exploring the advancements in Optimus.

By the end of this book, you'll be able to easily improve your data science workflow with Optimus.

Who this book is for

This book is for Python developers who want to explore, transform, and prepare big data for machine learning, analytics, and reporting using Spark and Dask on CPUs or GPUs. Beginner-level working knowledge of Python is assumed.

What this book covers

Chapter 1, Hi Optimus!, shows us what Optimus is, why it was created, and the goals of the project. We will get a good understanding of how Optimus works internally, how it is different from the current technology, and how it works in tandem with users to bring the best of the technology to serve users' data processing needs.

Chapter 2, Data Loading, Saving, and File Formats, is all about how to use Optimus to load and save data from text data formats such as CSV and JSON files. Also, we will explore binary files such as Excel and some optimized for columnar data processing such as Parquet, Avro, and OCR. Lastly, we will learn how to connect to databases such as SQLite and remote data storage such as Redshift.

Chapter 3, Data Wrangling, demonstrates how to concatenate data row- and column-wise and how to use SQL-like syntax to merge data using left, right, inner, and outer methods. Also, we'll learn how to pivot data tables to put data in the shape needed for the next step in the data pipeline.

Chapter 4, Combining, Reshaping, and Aggregating Data, teaches us how to group columns of data and apply summary statistics to each group of data. From count, min, and max aggregation to more advanced stats such as kurtosis and skew, we'll have all the tools to calculate any stats needed.

Chapter 5, Data Visualization and Profiling, demonstrates the profiler and the profiler data types, an Optimus DataFrame unique feature that lets the user have an overview of the data quality on every column. Data types such as email, dates, URLs, string, and float let the user easily standardize mismatches and missing data like any other library.

Chapter 6, String Clustering, uses string clustering techniques to let us easily identify groups of similar strings and replace them with a unique value.

Chapter 7, Feature Engineering, teaches us how to create new features for machine learning models to learn from. We'll look at generating them by combining fields, extracting values from messy columns, or encoding them for better results.

Chapter 8, Machine Learning, shows us how to easily create machine learning models and how Optimus will take care of the implementation details and make the feature engineering work when possible, as well as how to save the model after training and load it for future use.

Chapter 9, Natural Language Processing, shows us how to easily prepare data to apply techniques such as a word cloud, data summarization, and sentiment analysis, along with examples.

Chapter 10, Hacking Optimus, explores how to add new profiler data types to better approach quality problems. Also, we will learn how to create data cleaning and transformation functions and how to add a hypothetical new data processing engine. To close, we will talk about the Optimus Community, how to contribute to the project, and the next step in the Optimus project.

Chapter 11, Optimus as a Web Service, demonstrates how Optimus can be used as a web service with the help of various tools and plugins.

To get the most out of this book

Software/hardware covered in the book	Operating system requirements
JupyterLab	Windows, macOS, or Linux
Python 3.6 or above	Windows, macOS, or Linux
Optimus	Windows, macOS, or Linux

If you are using the digital version of this book, we advise you to type the code yourself or access the code from the book's GitHub repository (a link is available in the next section). Doing so will help you avoid any potential errors related to the copying and pasting of code.

Download the example code files

You can download the example code files for this book from GitHub at https://github.com/PacktPublishing/Data-Processing-with-Optimus. If there's an update to the code, it will be updated in the GitHub repository.

We also have other code bundles from our rich catalog of books and videos available at https://github.com/PacktPublishing/. Check them out!

Download the color images

We also provide a PDF file that has color images of the screenshots and diagrams used in this book. You can download it here: https://static.packt-cdn.com/downloads/9781801079563_ColorImages.pdf

Conventions used

There are a number of text conventions used throughout this book.

`Code in text`: Indicates code words in text, database table names, folder names, filenames, file extensions, pathnames, dummy URLs, user input, and Twitter handles. Here is an example: "For example, calling `df.display()` after any delayed function will require the final data to be calculated."

A block of code is set as follows:

```
from optimus import Optimus
op = Optimus("dask")
df = op.create.dataframe({"A":[0,1,2,3,4,5]})
df = df.cols.sqrt("A")
```

When we wish to draw your attention to a particular part of a code block, the relevant lines or items are set in bold:

```
from optimus import Optimus
op = Optimus("dask")
df = op.create.dataframe({"A":[0,1,2,3,4,5]})
df = df.cols.sqrt("A")
```

Any command-line input or output is written as follows:

```
coiled install optimus/default
conda activate coiled-optimus-default
```

Bold: Indicates a new term, an important word, or words that you see onscreen. For instance, words in menus or dialog boxes appear in **bold**. Here is an example: Enable WSL 2 by enabling the **Virtual Machine Platform** optional feature.

> **Tips or important notes**
> Appear like this.

Get in touch

Feedback from our readers is always welcome.

General feedback: If you have questions about any aspect of this book, email us at customercare@packtpub.com and mention the book title in the subject of your message.

Errata: Although we have taken every care to ensure the accuracy of our content, mistakes do happen. If you have found a mistake in this book, we would be grateful if you would report this to us. Please visit www.packtpub.com/support/errata and fill in the form.

Piracy: If you come across any illegal copies of our works in any form on the internet, we would be grateful if you would provide us with the location address or website name. Please contact us at copyright@packt.com with a link to the material.

If you are interested in becoming an author: If there is a topic that you have expertise in and you are interested in either writing or contributing to a book, please visit authors. packtpub.com.

Share Your Thoughts

Once you've read *Data Processing with Optimus*, we'd love to hear your thoughts! Scan the QR code below to go straight to the Amazon review page for this book and share your feedback.

https://packt.link/r/1-801-07956-0

Your review is important to us and the tech community and will help us make sure we're delivering excellent quality content.

Section 1: Getting Started with Optimus

By the end of this section, you will have a good understanding of what Optimus brings to the table in terms of improving the entire data processing landscape.

This section comprises the following chapters:

1
Hi Optimus!

Optimus is a Python library that loads, transforms, and saves data, and also focuses on wrangling tabular data. It provides functions that were designed specially to make this job easier for you; it can use multiple engines as backends, such as pandas, **cuDF**, **Spark**, and **Dask**, so that you can process both small and big data efficiently.

Optimus is *not* a DataFrame technology: it is not a new way to organize data in memory, such as **arrow**, or a way to handle data in GPUs, such as **cuDF**. Instead, Optimus relies on these technologies to load, process, explore, and save data.

Having said that, this book is for everyone, mostly data and machine learning engineers, who want to simplify writing code for data processing tasks. It doesn't matter if you want to process small or big data, on your laptop or in a remote cluster, if you want to load data from a database or from remote storage – Optimus provides all the tools you will need to make your data processing task easier.

In this chapter, we will learn about how Optimus was born and all the DataFrame technologies you can use as backends to process data. Then, we will learn about what features separate Optimus from all the various kinds of DataFrame technologies. After that, we will install **Optimus** and **Jupyter Lab** so that we will be prepared to code in *Chapter 2, Data Loading, Saving, and File Formats*.

Finally, we will analyze some of Optimus's internal functions to understand how it works and how you can take advantage of some of the more advanced features.

A key point: this book will not try to explain how every DataFrame technology works. There are plenty of resources on the internet that explain the internals and the day-to-day use of these technologies. Optimus is the result of an attempt to create an expressive and easy to use data API and give the user most of the tools they need to complete the data preparation process in the easiest way possible.

The topics we will be covering in this chapter are as follows:

- Introducing Optimus
- Installing everything you need to run Optimus
- Using Optimus
- Discovering Optimus internals

Technical requirements

To take full advantage of this chapter, please ensure you implement everything specified in this section.

Optimus can work with multiple backend technologies to process data, including GPUs. For GPUs, Optimus uses **RAPIDS**, which needs an NVIDIA card. For more information about the requirements, please go to the GPU configuration section.

To use RAPIDS on Windows 10, you will need the following:

- Windows 10 version 2004 (OS build 202001.1000 or later)
- CUDA version 455.41 in CUDA SDK v11.1

You can find all the code for this chapter in `https://github.com/PacktPublishing/Data-Processing-with-Optimus`.

Introducing Optimus

Development of Optimus began with work being conducted for another project. In 2016, Alberto Bonsanto, Hugo Reyes, and I had an ongoing big data project for a national retail business in Venezuela. We learned how to use **PySpark** and **Trifacta** to prepare and clean data and to find buying patterns.

But problems soon arose for both technologies: the data had different category/product names over the years, a 10-level categorization tree, and came from different sources, including CSV files, Excel files, and databases, which added an extra process to our workflow and could not be easily wrangled. On the other hand, when we tried Trifacta, we needed to learn its unique syntax. It also lacked some features we needed, such as the ability to remove a single character from every column in the dataset. In addition to that, the tool was closed source.

We thought we could do better. We wanted to write an open source, user-friendly library in Python that would let any non-experienced user apply functions to clean, prepare, and plot big data using PySpark.

From this, Optimus was born.

After that, we integrated other technologies. The first one we wanted to include was cuDF, which supports processing data 20x faster; soon after that, we also integrated Dask, Dask-cuDF, and Ibis. You may be wondering, why so many DataFrame technologies? To answer that, we need to understand a little bit more about how each one works.

Exploring the DataFrame technologies

There are many different well-known DataFrame technologies available today. Optimus can process data using one or many of those available technologies, including pandas, Dask, cuDF, Dask-cuDF, Spark, Vaex, or Ibis.

Let's look at some of the ones that work with Optimus:

- **pandas** is, without a doubt, one of the more popular DataFrame technologies. If you work with data in Python, you probably use pandas a lot, but it has an important caveat: pandas cannot handle multi-core processing. This means that you cannot use all the power that modern CPUs can give you, which means you need to find a hacky way to use all the cores with pandas. Also, you cannot process data volumes greater than the memory available in RAM, so you need to write code to process your data in chunks.

- **Dask** came out to help parallelize Python data processing. In Dask, we have the Dask DataFrame, an equivalent to the pandas DataFrame, that can be processed in parallel using multiple cores, as well as with nodes in a cluster. This gives the user the power to scale out data processing to hundreds of machines in a cluster. You can start 100 machines, process your data, and shut them down, quickly, easily, and cheaply. Also, it supports out-of-core processing, which means that it can process data volumes greater than the memory available in RAM.

- At the user level, **cuDF** and **Dask-cuDF** work in almost the same way as pandas and Dask, but up to 20x faster for most operations when using GPUs. Although GPUs are expensive, they give you more value for money compared to CPUs because they can process data much faster.

- **Vaex** is growing in relevance in the DataFrame landscape. It can process data out-of-core, is easier to use than Dask and PySpark, and is optimized to process string and stats in parallel because of its underlying C language implementation.

- **Ibis** is gaining traction too. The amazing thing about Ibis is that it can use multiple engines (like Optimus but focused on SQL engines) and you can write code in Python that can be translated into SQL to be used in Impala, MySQL, PostgreSQL, and so on.

The following table provides a quick-glance comparison of several of these technologies:

Dataframe Technology	CPU	GPU	Out-of-Core	Distributed Dataframe (Multi-Core/Multi-None)
Pandas	x			
cuDF				
Dask	x	x	x	x
Dask-cuDF		x	x	x
PySpark	x	x	x	x
Vaex		x	x	Work in Progress
Ibis	x			

(*) Depends on the engine that's been configured

Figure 1.1 – DataFrame technologies and capabilities available in Optimus

There are some clear guidelines regarding when to use each engine:

- Use pandas if the DataFrame fits comfortably in memory, or cuDF if you have GPUs and the data fits in memory. This is almost always faster than using distributed DataFrame technologies under the same circumstances. This works best for real-time or near-real-time data processing.

- Use Dask if you need to process data greater than memory, and Dask-cuDF if you have data larger than memory and a multi-core and/or multi-node GPU infrastructure.

- Use Vaex if you have a single machine and data larger than memory, or Spark if you need to process data at terabyte scale. This is slow for small datasets/datasets that fit in memory.

Now that you understand this, you can unleash Optimus's magic and start preparing data using the same Optimus API in any of the engines available.

Examining Optimus design principles

A key point about Optimus is that we are not trying to create a new DataFrame technology. As we've already seen, there are actually many amazing options that cover almost any use case. The purpose of Optimus is to simplify how users handle data and give that power to people who may not have any technical expertise. For that reason, Optimus follows three principles:

- *One API to rule them all.*

- *Knowing the technology is optional.*

- *Data types should be as rich as possible.*

What do these mean? Let's look at each in detail.

One API to rule them all

Almost all DataFrame technologies try to mimic the pandas API. However, there are subtle differences regarding what the same function can do, depending on how you apply it; with Optimus, we want to abstract all this.

We'll go into more detail about this later, but here's a quick example: you can calculate a column square root using the .cols accessor, like so:

```
from optimus import Optimus
op = Optimus("dask")
df = op.create.dataframe({"A":[0,-1,2,3,4,5]})
df = df.cols.sqrt("A")
```

If you want to switch from Dask to any other engine, you can use any of the following values. Each one will instantiate a different class of the Optimus DataFrame:

- `"pandas"` to use Pandas. This will instantiate a pandas DataFrame.
- `"dask"` to use Dask. This will instantiate a DaskDataFrame.
- `"cudf"` to use cuDF. This will instantiate a CUDFDataFrame.
- `"dask_cudf"` to use Dask-cuDF. This will instantiate a DaskCUDFDataFrame.
- `"spark"` to use PySpark. This will instantiate a SparkDataFrame.
- `"vaex"` to use Vaex. This will instantiate a VaexDataFrame.
- `"ibis"` to use Ibis. This will instantiate an IbisDataFrame.

An amazing thing about this flexibility is that you can process a sample of the data on your laptop using pandas, and then send a job to Dask-cuDF or a Spark cluster to process it using the faster engine.

Knowing the technical details is optional

pandas is complex. Users need to handle technical details such as rows, index, series, and masks, and you need to go low level and use **NumPy/Numba** to get all the power from your CPU/GPU.

With Numba, users can gain serious speed improvements when processing numerical data. It translates Python functions into optimized machine code at runtime. This simply means that we can write faster functions on CPU or GPU. For example, when we request a histogram using Optimus, the minimum and maximum values of a column are calculated in a single pass.

In Optimus, we try to take the faster approach for every operation, without requiring extensive knowledge of the technical details, to take full advantage of the technology. That is Optimus's job.

Some other DataFrame features that are abstracted in Optimus include indices, series, and masks (the exception is PySpark). In Optimus, you only have columns and rows; the intention is to use familiar concepts from spreadsheets so that you can have a soft landing when you start using Optimus.

In Optimus, you have two main accessors, `.cols` and `.rows`, that provide most of the transformation methods that are available. For example, you can use `df.cols.lower` to transform all the values of a column into lowercase, while you can use `df.rows.drop_duplicates` to drop duplicated rows in the entire dataset. Examples of these will be addressed later in this book.

Data types should be as rich as possible

All DataFrame technologies have data types to represent integers, decimals, time, and dates. In pandas and Dask, you can use NumPy data types to assign diverse types of integers such as int8, int16, int32, and int64, or different decimals types, such as float32, float64, and so on.

This gives the user a lot of control to optimize how the data is saved and reduces the total size of data in memory and on disk. For example, if you have 1 million rows with values between 1 and 10, you can save the data as uint8 instead of inf64 to reduce the data size.

Besides this internal data representation, Optimus can infer and detect a richer set of data types so that you can understand what data in a column matches a specific data type (URL, email, date, and so on) and then apply specific functions to handle it.

In Optimus, we use the term **quality** to express three data characteristics:

- Number of values that match the data type being inferred
- Number of values that differ from the data type being inferred
- Number of missing values

Using the df.cols.quality method, Optimus can infer the data type of every loaded column and return how many values in the column match its data types. In the case of **date** data types, Optimus can infer the date format.

The following list shows all the data types that Optimus can detect:

- Integer.
- Strings.
- Email.
- URL.
- Gender.
- Boolean.
- US ZIP code.
- Credit card.
- Time and date format.
- Object.
- Array.

- Phone number.
- Social security number.
- HTTP code.

Many of these data types have special functions that you can use, as follows:

- URL: Schemas, domains, extensions, and query strings
- Date: Year, month, and day
- Time: Hours, minutes, and seconds
- Email: domains and domain extensions

The best part of this is that you can define your own data types and create special functions for them. We will learn more about this later in this book. We will also learn about the functions we can use to process or remove matches, mismatches, and missing values.

Now that we've had a look at how Optimus works, let's get it running on your machine.

Installing everything you need to run Optimus

To start using Optimus, you will need a laptop with Windows, Ubuntu, or macOS installed with support for Python, **PIP packages**, and **Conda**. If you are new to Python, PIP is the main package manager. It allows Python users to install and manage packages that expand the Python standard library's functionality.

The easiest way to install Optimus is through PIP. It will allow us to start running examples in just a few minutes. Later in this section, we will see some examples of Optimus running on a notebook, on a shell terminal, and on a file read by Python, but before that, we will need to install Optimus and its dependencies.

First, let's install **Anaconda**.

Installing Anaconda

Anaconda is a free and open source distribution of the Python and R programming languages. The distribution comes with the Python interpreter, Conda, and various packages related to machine learning and data science so that you can start easier and faster.

To install Anaconda on any system, you can use an installer or install it through a system package manager. In the case of Linux and macOS, you can install Anaconda using APT or Homebrew, respectively.

On Linux, use the following command:

```
sudo apt-get install anaconda # on Linux
```

For macOS and Windows, go to `https://www.anaconda.com/products/individual`. Download the Windows file that best matches your system and double-click the file after downloading it to start the installation process:

```
brew cask install anaconda # on macOS
```

With Anaconda now installed, let's install Optimus.

Installing Optimus

With Anaconda installed, we can use Conda to install Optimus:

As stated on the Conda website, Conda is provides *"package, dependency, and environment management for any language."* With Conda, we can manage multiple Python environments without polluting our system with dependencies. For example, you could create a Conda environment that uses Python 3.8 and pandas 0.25, and another with Python 3.7 and pandas 1.0. Let's take a look:

1. To start, we need to open the Anaconda Prompt. This is just the command-line interface that comes with Conda:

   - *For Windows*: From the Start menu, search for and open **Anaconda Prompt**.
   - *For macOS*: Open Launchpad and click the Terminal icon.
   - *For Linux*: Open a Terminal window.

2. Now, in the terminal, we are going to create a Conda environment named Optimus to create a clean Optimus installation:

   ```
   conda create -n optimus python=3.8
   ```

3. Now, you need to change from the (base) environment to the (optimus) environment using the following command:

   ```
   conda activate optimus
   ```

4. Running the following command on your terminal will install Optimus with its basic features, ready to be tested:

```
pip install pyoptimus
```

5. If you have done this correctly, running a simple test will tell us that everything is correct:

```
python -c 'import optimus; optimus.__version__'
```

Now, we are ready to use Optimus!

We recommend using **Jupyter Notebook**, too.

Installing JupyterLab

If you have not been living under a rock the last 5 years, you probably know about Jupyter Notebook. **JupyterLab** is the next generation of Jupyter Notebook: it is a web-based interactive development environment for coding. Jupyter (for short) will help us test and modify our code easily and will help us try out our code faster. Let's take a look:

1. To install JupyterLab, go to the Terminal, as explained in the *Installing Optimus* section, and run the following command:

```
conda install -c conda-forge jupyterlab
```

2. At this point, you could simply run Jupyter. However, we are going to install a couple of handy extensions to debug Dask and track down GPU utilization and RAM:

```
conda install nodejs
conda install -c conda-forge dask-labextension
jupyter labextension install dask-labextension
jupyter serverextension enable dask_labextension
```

3. Now, let's run Jupyter using the following command:

```
jupyter lab --ip=0.0.0.0
```

4.   You can access Jupyter using any browser:

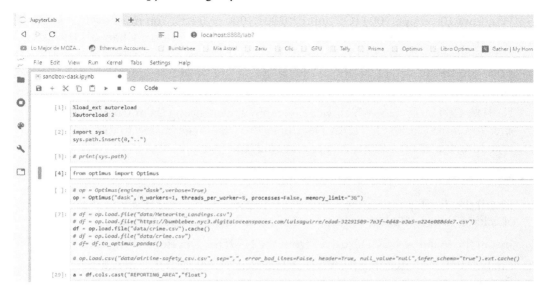

Figure 1.2 – JupyterLab UI

Next, let's look at how to install RAPIDS.

Installing RAPIDS

There are some extra steps you must take if you want to use a GPU engine with Optimus.

RAPIDS is a set of libraries developed by NVIDIA for handling end-to-end data science pipelines using GPUs; cuDF and Dask-cuDF are among these libraries. Optimus can use both to process data in a local and distributed way.

For RAPIDS to work, you will need a GPU, NVIDIA Pascal™ or better, with a compute capability of 6.0+. You can check the compute capability by looking at the tables on the NVIDIA website: `bit.ly/cc_gc`.

First, let's install RAPIDS on Windows.

Installing RAPIDS on Windows 10

RAPIDS is not fully supported at the time of writing (December 2020), so you must use the **Windows Subsystem for Linux version 2 (WSL2)**. WSL is a Windows 10 feature that enables you to run native Linux command-line tools directly on Windows.

You will need the following:

- Windows 10 version 2004 (OS build 202001.1000 or later). You must sign up to get Windows Insider Preview versions, specifically to the Developer Channel. This is required for the WSL2 VM to have GPU access: `https://insider.windows.com/en-us/`.

- CUDA version 455.41 in CUDA SDK v11.1. You must use a special version of the NVIDA CUDA drivers, which you can get by downloading them from NVIDIA's site. You must join the NVIDIA Developer Program to get access to the version; searching for `WSL2 CUDA Driver` should lead you to it.

Here are the steps:

1. Install the developer preview version of Windows. Make sure that you click the checkbox next to **Update** to install other recommended updates too.
2. Install the Windows CUDA driver from the NVIDIA Developer Program.
3. Enable WSL 2 by enabling the **Virtual Machine Platform** optional feature. You can find more steps here: `https://docs.microsoft.com/en-us/windows/wsl/install-win10`.
4. Install WSL from the Windows Store (Ubuntu-20.04 is confirmed to be working).
5. Install Python on the WSL VM, tested with Anaconda.
6. Go to the *Installing RAPIDS* section of this chapter.

Installing RAPIDS on Linux

First, you need to install the CUDA and NVIDIA drivers. Pay special attention if your machine is running code that depends on a specific CUDA version. For more information about the compatibility between the CUDA and NVIDIA drivers, check out `bit.ly/cuda_c`.

If you do not have a compatible GPU, you can use a cloud provider such as Google Cloud Platform, Amazon, or Azure.

In this case, we are going to use Google Cloud Platform. As of December 2020, you can get an account with 300 USD on it to use. After creating an account, you can set up a VM instance to install RAPIDS.

To create a VM instance on Google Cloud Platform, follow these steps:

1. First, go to the hamburger menu, click **Compute Engine**, and select **VM Instances**.

2. Click on **CREATE INSTANCE**. You will be presented with a screen that looks like this:

Figure 1.3 – Google Cloud Platform instance creation

3. Select a region that can provide a GPU. Not all zones have GPUs available. For a full list, check out `https://cloud.google.com/compute/docs/gpus`.

4. Make sure you choose **N1** series from the dropdown.

5. Be sure to select an OS that's compatible with the CUDA drivers (check the options available here: `https://developer.nvidia.com/cuda-downloads`). After the installation, you will be using 30 GB of storage space, so make sure you assign enough disk space:

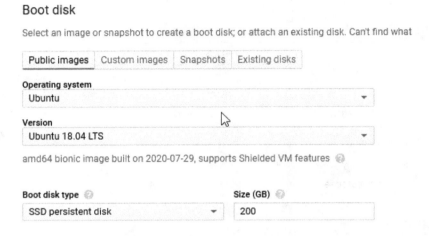

Figure 1.4 – Google Cloud Platform OS selection

6. Check the Allow **HTTP traffic** option:

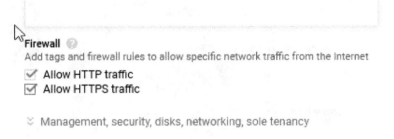

Figure 1.5 – Google Cloud Platform OS selection

7. To finish, click the **Create** button at the bottom of the page:

Figure 1.6 – Google Cloud instance creation

Now, you are ready to install RAPIDS.

Installing RAPIDS

After checking that your GPU works with Optimus, go to `https://rapids.ai/start.html`. Select the options that match your requirements and copy the output from the command section to your command-line interface:

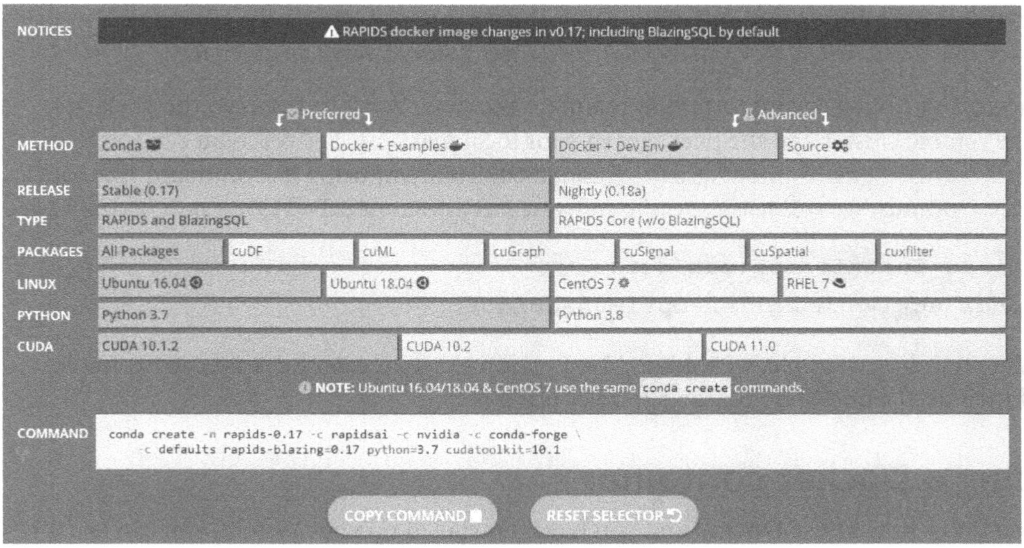

Figure 1.7 – Google Cloud Platform OS selection

After the installation process is complete, you can test RAPIDS by importing the library and getting its version:

```
python -c 'import cudf; cudf.__version__'
```

Next, let's learn how to install **Coiled** for easier setups.

Using Coiled

Coiled is a deployment-as-a-service library for scaling Python that facilitates Dask and Dask-cuDF clusters for users. It takes the DevOps out of the data role to enable data professionals to spend less time setting up networking, managing fleets of Docker images, creating AWS IAM roles, and other setups they would have to handle otherwise, so that they can spend more time on their real job.

To use a Coiled cluster on Optimus, we can just pass minimal configuration to our Optimus initialization function and include our token provided by Coiled in a parameter; to get this token, you must create an account at `https://cloud.coiled.io` and get the token from your dashboard, like so:

```
op = Optimus(coiled_token="<your token here>", n_workers=2)
```

In this example, we initialized Optimus using a Coiled token, and set the number of workers to 2. Optimus will initialize a Dask DataFrame and handle the connection to the cluster that was created by Coiled. After this, Optimus will work as normal.

When using Coiled, it's important to maintain the same versions between the packages in the remote cluster and the packages in your local machine. For this, you can install a Coiled software environment as a local conda environment using its command-line tool. To use Optimus, we will use a specific software environment called `optimus/default`:

```
coiled install optimus/default
conda activate coiled-optimus-default
```

In the preceding example, we told `coiled install` to create the conda environment and then used `conda activate` to start using it.

Using a Docker container

If you know how to use **Docker** and you have it installed on your system, you can use it to quickly set up Optimus in a working environment.

To use Optimus in a Docker environment, simply run the following command:

```
docker run -p 8888:8888 --network="host" optimus-df/
optimus:latest
```

This will pull the latest version of the Optimus image from Docker Hub and run a notebook process inside it. You will see something like the following:

```
To access the notebook, open this file in a browser:
    file://...
Or copy and paste one of these URLs:
    http://127.0.0.1:8888/?token=<GENERATED TOKEN>
```

Just copy the address and paste it into your browser, making sure it has the same token, and you'll be using a notebook with Optimus installed in its environment.

Using Optimus

Now that we have Optimus installed, we can start using it. In this section, we'll run through some of the main features and how to use them.

The Optimus instance

You use the Optimus instance to configure the engine, as well as load and save data. Let's see how this works.

Once Optimus has been installed on your system, you can use it in a Python environment. Let's import the Optimus class and instantiate an object of it:

```
from optimus import Optimus
op = Optimus(engine=pandas)
```

In Optimus, we call a DataFrame technology an engine. In the preceding example, we're setting up Optimus using Pandas as the base engine. Very easy!

Now, let's instantiate Optimus using Dask in a remote cluster. For this, we'll have to pass the configuration in the arguments to the Optimus function – specifically, the session argument – which allows us to pass a Dask client:

```
from dask.distributed import Client
client = Client("127.0.0.105")
op = Optimus(engine="dask", session=client)
```

In the preceding code, we instantiated a Dask distributed client and passed it to the Optimus initialization.

To initialize with a different number of workers, you can pass a named argument as well:

```
op = Optimus(engine="dask", n_workers=2)
```

This will create a local client automatically, instead of passing just one, as in the previous example.

Using Dask, you can now access more than 100 functions to transform strings, as well as filter and merge data.

Saving and loading data from any source

Using the Optimus instance, you can easily load DataFrames from files or databases. To load a file, simply call one of the available methods for different formats (`.csv`, `.parquet`, `.xlsx`, and more) or simply the generic `file` method, which will infer the file format and other parameters:

```
op.load.csv("path/to/file.csv")
op.load.file("path/to/file.csv")
```

For databases or external buckets, Optimus can handle connections as different instances, which allows us to maintain operations and clean any credentials and addresses that may or may not repeat on different loading routines:

```
db = op.connect.database( *db_args )
op.load.database_table("table name", connection=db)
conn = op.connect.s3( *s3_args )
op.load.file("relative/path/to/file.csv", connection=conn)
```

On the other hand, to save to a file or to the table of a database, you can use the following code:

```
df.save.csv("relative/path/to/file.csv", connection=conn)
df.save.database_table("table_name", db=db)
```

Now that we have started our engine and have our data ready, let's see how we can process it using the Optimus DataFrame.

The Optimus DataFrame

One of the main goals of Optimus is to try and provide an understandable, easy-to-remember API, along with all the tools needed to clean and shape your data. In this section, we are going to highlight the main features that separate Optimus from the available DataFrame technologies.

Using accessors

Optimus DataFrames are made to be modified in a natural language, dividing the different methods available into components. For example, if we want to modify a whole column, we may use the methods available in the `.cols` accessor, while if we want to filter rows by the value of a specific column, we may use a method in the `.rows` accessor, and so on.

An example of an operation could be column renaming:

```
df.cols.rename("function", "job")
```

In this case, we are simply renaming the function column to "job", but the modified DataFrame is not saved anywhere, so the right way to do this is like so:

```
df = df.cols.rename("function", "job")
```

In addition, most operations return a modified version of the DataFrame so that those methods can be called, chaining them:

```
df = df.cols.upper("name").cols.lower("job")
```

When Optimus instantiates a DataFrame, it makes an abstraction of the core DataFrame that was made using the selected engine. There is a DataFrame class for every engine that's supported. For example, a Dask DataFrame is saved in a class called DaskDataFrame that contains all the implementations from Optimus. Details about the internals of this will be addressed later in this book.

As we mentioned previously, to use most of these methods on an Optimus DataFrame, it's necessary to use accessors, which separate different methods that may have distinct behaviors, depending on where they're called:

```
df.cols.drop("name")
```

The preceding code will drop the entire "name" column. The following command returns a different DataFrame:

```
df.rows.drop(df["name"]==MEGATRON)
```

The preceding code will drop the rows with values in the "name" column that match the MEGATRON value.

Obtaining richer DataFrame data

Optimus aims to give the user important information when needed. Commonly, you will use head() or show() to print DataFrame information. Optimus can provide additional useful information when you use display:

```
df.display()
```

This produces the following output:

N/A partition(s)

name	**function**
1 (object)	2 (object)
not nullable	not nullable
OPTIMUS	·LEADER
BUMBLEBEE	·ESPIONAGE
EJECT	·ELECTRONIC·SURVEILLANCE

Viewing 3 of 3 rows / 2 columns

N/A partition(s) <class 'optimus.engines.pandas.dataframe.PandasDataFrame'>

Figure 1.8 – Optimus DataFrame display example

In the previous screenshot, we can see information about the requested DataFrame, such as the number of columns and rows, all the columns, along with their respective data types, some values, as well as the number of partitions and the type of the queried DataFrame (in this case, a DaskDataFrame). This information is useful when you're transforming data to make sure you are on the right track.

Automatic casting when operating

Optimus will cast the data type of a column based on the operation you apply. For example, to calculate min and max values, Optimus will convert the column into a float and ignore non-numeric data:

```
dfn = op.create.dataframe({"A":["1",2,"4","!",None]})
dfn.cols.min("A"), df.cols.max("A")
(1.0, 4.0)
```

In the preceding example, Optimus ignores the "!" and None values and only returns the lower and higher numeric values, respectively, which in this case are 1.0 and 4.0.

Managing output columns

For most column methods, we can choose to make a copy of every input column and save it to another so that the operation does not modify the original column. For example, to save an uppercase copy of a string type column, we just need to call the same df.cols. upper with an extra argument called output_cols:

```
df.cols.capitalize("name", output_cols="cap_name")
```

This parameter is called output_cols and is plural because it supports multiple names when multiple input columns are passed to the method:

```
df.cols.upper(["a", "b"],
              output_cols=["upper_a", "upper_b"])
```

In the preceding example, we doubled the number of columns in the resulting DataFrame, one pair untouched and another pair with its values transformed into uppercase.

Profiling

To get an insight into the data being transformed by Optimus, we can use df.profile(), which provides useful information in the form of a Python dictionary:

```
df = op.create.dataframe({"A":["1",2,"4","!",None],
                          "B":["Optimus","Bumblebee",
                               "Eject", None, None]})
df.profile(bins=10)
```

This produces the following output:

```
{'columns': {'A': {'stats': {'match': 0,
    'missing': 1,
    'mismatch': 0,
    'profiler_dtype': {'dtype': 'int'},
    'hist': [{'lower': 1.0, 'upper': 1.3333333333333333, 'count': 1},
    {'lower': 1.3333333333333333, 'upper': 1.6666666666666665, 'count': 0},
    {'lower': 1.6666666666666665, 'upper': 2.0, 'count': 0},
    {'lower': 2.0, 'upper': 2.3333333333333333, 'count': 1},
    {'lower': 2.3333333333333333, 'upper': 2.6666666666666665, 'count': 0},
    {'lower': 2.6666666666666665, 'upper': 3.0, 'count': 0},
    {'lower': 3.0, 'upper': 3.3333333333333333, 'count': 0},
    {'lower': 3.3333333333333333, 'upper': 3.6666666666666665, 'count': 0},
    {'lower': 3.6666666666666665, 'upper': 4.0, 'count': 1}],
    'count_uniques': 5},
   'dtype': 'object'},
  'B': {'stats': {'match': 3,
    'missing': 2,
    'mismatch': 2,
    'profiler_dtype': {'dtype': 'string'},
    'frequency': [{'value': 'Bumblebee', 'count': 1},
    {'value': 'Optimus', 'count': 1},
    {'value': 'Eject', 'count': 1}],
    'count_uniques': 3},
   'dtype': 'object'}},
 'name': None,
 'file_name': None,
 'summary': {'cols_count': 2,
  'rows_count': 5,
  'dtypes_list': ['object'],
  'total_count_dtypes': 1,
  'missing_count': 0,
  'p_missing': 0.0}}
```

Figure 1.9 – Profiler output

In the preceding screenshot, we can see data, such as every column, along with its name, missing values, and mismatch values, its inferred data type, its internal data type, a histogram (or values by frequency in categorical columns), and unique values. For the DataFrame, we have the name of the DataFrame, the name of the file (if the data comes from one), how many columns and rows we have, the total count of data types, how many missing rows we have, and the percentage of values that are missing.

Visualization

One of the most useful features of Optimus is its ability to plot DataFrames in a variety of visual forms, including the following:

- Frequency charts
- Histograms

- Boxplots
- Scatterplots

To achieve this, Optimus uses Matplotlib and Seaborn, but you can also get the necessary data in Python Dictionary format to use with any other plotting library or service.

Python Dictionary output

By default, every output operation in Optimus will get us a dictionary (except for some cases, such as aggregations, which get us another DataFrame by default). Dictionaries can easily be transformed and saved into a JSON file, in case they are needed for a report or to provide data to an API:

```
df.columns_sample("*")
```

String, numeric, and encoding tools

Optimus tries to provide out-of-the-box tools to process strings and numbers, and gives you tools for the data preparation process so that you can create machine learning models.

String clustering

String clustering refers to the operation of grouping different values that might be alternative representations of the same thing. A good example of this are the strings "NYC" and "New York City". In this case, they refer to the same thing.

Processing outliers

Finding outliers is one of the most common applications for statistical DataFrames. When finding outliers, there are various possible methods that will get us different results, such as **z-score**, **fingerprint**, **n-gram fingerprint**, and more. With Optimus, these methods are provided as alternatives so that you can adapt to most cases.

Encoding techniques

Encoding is useful for machine learning models since they require all the data going out or coming in to be numeric. In Optimus, we have methods such as string_to_index, which allows us to transform categorical data into numerical data.

Technical details

When dealing with distributed DataFrame technologies, there are two concepts that arise that are an integral part of how Optimus is designed. These are **lazy** and **eager** execution.

Let's explore how this works in Optimus.

Distributed engines process data in a lazy way

In Dask, for example, when you apply a function to a DataFrame, it is not applied immediately like it would be in pandas. You need to trigger this computation explicitly, by calling `df.execute()`, or implicitly, when calling other operations that trigger this processing.

Optimus makes use of this functionality to trigger all the computation when explicitly requested. For example, when we request the profile of a dataset, every operation, such as **histogram calculation**, **top-n values**, and **data types inference**, is pushed to a cluster in a **directed acyclic graph** (**DAG**) and executed in parallel.

The following representation shows multiple operations being executed in parallel being visualized in a DAG:

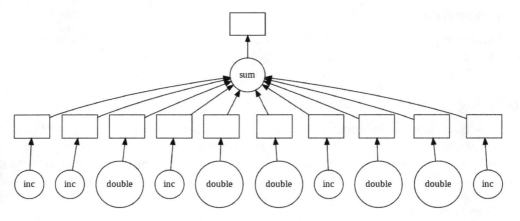

Figure 1.10 – Visualizing a DAG in Dask

Aggregations return computed operations (eager execution)

As we mentioned earlier, distributed engines process aggregations in a lazy way. Optimus triggers aggregation so that you can always visualize the result immediately.

Triggering an execution

Optimus is capable of immediately executing an operation if requested. This only applies to engines that support delayed operations, such as Dask and PySpark. This way, we reduce the computation time when you know some operations will change on your workflow:

```
df = df.cols.replace("address", "MARS PLANET",
                        "mars").execute()
```

In the preceding example, we're replacing every match with `"MARS PLANET"` on the `address` column with `"mars"` and then saving a cache of this operation.

However, there are some operations or methods that will also trigger all the delayed operations that were made previously. Let's look at some of these cases:

- *Requesting a sample*: For example, calling `df.display()` after any delayed function will require the final data to be calculated. For this reason, Optimus will trigger all the delayed operations before requesting any kind of output; this will also happen when we call any other output function.

- *Requesting a profile*: When calling `df.profile()`, some aggregations are done in the background, such as counting for unique, mismatched, and missing values. Also, getting the frequent values or the histogram calculation of every column will require a delayed function to have been executed previously.

When using a distributed DataFrame technology, when an operation is executed, the data is executed and stored in every worker. In the case of Dask, this function would be `cache`, which is called when we call `execute` on our Optimus DataFrame. Note that if we call `compute` directly on a Dask DataFrame instead, all the data will be brought to the client, which might cause memory issues in our system if the data is too big.

Discovering Optimus internals

Optimus is designed to be easy to use for non-technical users and developers. Once you know how some of the internals work, you'll know how some transformations work, and hopefully how to avoid any unexpected behavior. Also, you'll be able to expand Optimus or make more advanced or engine-specific transformations if the situation requires it.

Engines

Optimus handles all the details that are required to initialize any engine. Although **pandas**, **Vaex**, and **Ibis** won't handle many configuration parameters because they are non-distributed engines, Dask and Spark handle many configurations, some of which are mapped and some of which are passed via the `*args` or `**kwargs` arguments.

Optimus always keeps a reference to the engine you initialize. For example, if you want to get the Dask client from the Optimus instance, you can use the following command:

```
op.client
```

This will show you the following information:

Client	Cluster
Scheduler: inproc://192.168.86.249/12148/10	**Workers:** 1
Dashboard: http://192.168.86.249:9679/status	**Cores:** 8
	Memory: 4.00 GB

Figure 1.11 – Dask client object inside Optimus

One interesting thing about Optimus is that you can use multiple engines at the same time. This might seem weird at first, but it opens up amazing opportunities if you get creative. For example, you can combine Spark, to load data from a database, and pandas, to profile a data sample in real time, or use pandas to load data and use Ibis to output the instructions as a set of SQL instructions.

At the implementation level, all the engines inherit from `BaseEngine`. Let's wrap all the engine functionality to make three main operations:

- **Initialization**: Here, Optimus handles all the initialization processes for the engine you select.
- **Dataframe creation**: `op.create.dataframe` maps to the DataFrame's creation, depending on the engine that was selected.
- **Data Loading**: `op.load` handles file loading and databases.

The DataFrame behind the DataFrame

The Optimus DataFrame is a wrapper that exposes and implements a set of functions to process string and numerical data. Internally, when Optimus creates a DataFrame, it creates it using the engine you select to keep a reference in the `.data` property. The following is an example of this:

```
op = Optimus("pandas")
df = op.load.csv("foo.txt", sep=",")
type(df.data)
```

This produces the following result:

```
Pandas.core.frame.DataFrame
```

A key point is that Optimus always keeps the data representation as DataFrames and not as a Series. This is important because in pandas, for example, some operations return a Series as result.

In pandas, use the following code:

```
import pandas as pd
type(pd.DataFrame({"A":["A",2,3]})["A"].str.lower())
pandas.core.series.Series
```

In Optimus, we use the following code:

```
from optimus import Optimus
op = Optimus("pandas")
type(op.create.dataframe({"A":["A",2,3]}).cols.lower().data)
pandas.core.frame.DataFrame
```

As you can see, both values have the same types.

Meta

Meta is used to keep some data that does not belong in the core dataset, but can be useful for some operations, such as saving the result of a top-N operation in a specific column. To achieve this, we save metadata in our DataFrames. This can be accessed using `df.meta`. This metadata is used for three main reasons. Let's look at each of them.

Saving file information

If you're loading a DataFrame from a file, it saves the file path and filename, which can be useful for keeping track of the data being handled:

```
from optimus import Optimus
op = Optimus("pandas")
df = op.load.csv("foo.txt", sep=",")
df.meta
```

You will get the following output:

```
{'file_name': 'foo.txt', 'name': 'foo.txt'}
```

Data profiling

Data cleaning is an iterative process; maybe you want to calculate the histogram or top-N values in the dataset to spot some data that you want to remove or modify. When you calculate profiling for data using `df.profile()`, Optimus will calculate a histogram or frequency chart, depending on the data type. The idea is that while working with the `Actions` data, we can identify when the histogram or top-N values should be recalculated. Next, you will see how **Actions** work.

Actions

As we saw previously, Optimus tries to cache certain operations to ensure that you do not waste precious compute time rerunning tasks over data that has not changed.

To optimize the cache usage and reconstruction, Optimus handles multiple internal Actions to operate accordingly.

You can check how Actions are saved by trying out the following code:

```
from optimus import Optimus
op = Optimus("pandas")
df = op.load.csv("foo.txt", sep=",")
df = df.cols.upper("*")
```

To check the actions you applied to the DataFrame, use the following command:

```
df.meta["transformations"]
```

You will get a Python dictionary with the action name and the column that's been affected by the action:

```
{'actions': [[{'upper': ['name']}], [{'upper': ['function']}]]}
```

A key point is that different actions have different effects on how the data is profiled and how the DataFrame's metadata is handled. Every Optimus operation has a unique Action name associated with it. Let's look at the five Actions that are available in Optimus and what effect they have on the DataFrame:

- **Columns**: These actions are triggered when operations are applied to entire Optimus columns; for example, `df.cols.lower()` or `df.cols.sqrt()`.

- **Rows**: These actions are triggered when operations are applied to any row in an Optimus column; for example, `df.rows.set()` or `df.rows.drop_duplicate()`.

- **Copy**: Triggered only for a copy column operation, such as `df.cols.copy()`. Internally, it just *creates* a new key on the `dict` meta with the source metadata column. If you copy an Optimus column, a profiling operation is not triggered over it.

- **Rename**: Triggered only for a rename column operation, such as `df.cols.rename()`. Internally, it just renames a key in the meta dictionary. If you copy an Optimus column, a profiling operation is not triggered over it.

- **Drop**: Triggered only for a rename column operation, such as `df.cols.drop()`. Internally, it *removes* a key in the meta dictionary. If you copy an Optimus column, a profiling operation is not triggered over it.

Dummy functions

There are some functions that do not apply to all the DataFrame technologies. Functions such as `.repartition()`, `.cache()`, and `compute()` are used in distributed DataFrames such as Spark and Dask to trigger operations in the workers, but these concepts do not exist in pandas or cuDF. To preserve the API's cohesion in all the engines, we can simply use `pass` or return the same DataFrame object.

Diagnostics

When you use Dask and Spark as your Optimus engine, you have access to their respective diagnostics dashboards. For very complex workflows, it can be handy to understand what operations have been executed and what could be slowing down the whole process.

Let's look at how this works in the case of Dask. To gain access to the diagnostic panel, you can use the following command:

```
op.client()
```

This will provide you with information about the Dask client:

Client

Scheduler: inproc://192.168.86.249/4676/1
Dashboard: http://192.168.86.249:39011/status

Cluster

Workers: 1
Cores: 8
Memory: 4.00 GB

Figure 1.12 – Dask client information

In this case, you can point to http://192.168.86.249:39011/status in your browser to see the Dask Diagnostics dashboard:

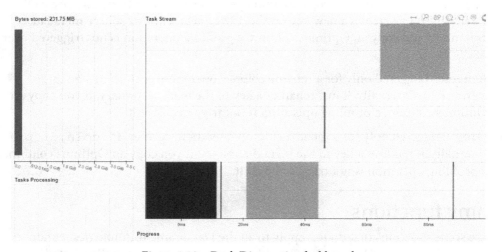

Figure 1.13 – Dask Diagnostics dashboard

An in-depth discussion about diagnostics is beyond the scope of this book. To find out more about this topic, go to https://docs.dask.org/en/latest/diagnostics-distributed.html.

Summary

In this chapter, we learned about Optimus's basic capabilities and which of the available engines is the most suitable, based on the infrastructure you're using. You also learned why it is beneficial to use Optimus instead of a vanilla DataFrame and what features separate Optimus from the DataFrame technology that's available. Then, we learned how to install Optimus on Windows, macOS, and Linux, both on the cloud and with external services such as **Coiled**.

Finally, we took a deep dive into the internals of Optimus so that you have a better understanding of how it works and how it allows you to get creative and take full advantage of Optimus.

In the next chapter, we will learn how to load and save data from files, databases, and remote locations such as Amazon S3.

2
Data Loading, Saving, and File Formats

Most of the data in the world is saved in files or databases, in local or remote sources. In this chapter, we will learn how to load data from multiple formats and data sources and how to save it, while looking at every method that can be used in detail.

Optimus puts a heavy focus on data sources that have been optimized for big data processing, such as **Avro**, **Parquet**, and **ORC**, and databases such as **BigQuery** and **Redshift**, so that users have all the tools they need at hand to cover their data processing needs.

From a developer's standpoint, Optimus follows the "batteries included" paradigm, so you don't have to worry about installing extra libraries to handle the **Excel** or **Avro** files that we use to handle this format. However, in the case of **databases**, every engine has their own respective driver, so including all the drivers inside a package would be problematic: it would be a package that's almost 500 MB in size!

The topics we will be covering in this chapter are as follows:

- How data moves internally
- Loading a file
- Creating a dataframe from scratch
- Connecting to remote data sources
- Saving a dataframe

Technical requirements

Optimus can work with multiple backend technologies to process data, including GPUs. For GPUs, Optimus use **RAPIDS**, which needs an NVIDIA card. For more information about the requirements, please go to the *GPU configuration* section of *Chapter 1*.

For databases, Optimus will need different database drivers to handle different technologies. In the *Connecting to databases* section, we will explain how to install every driver for every database technology.

You can find all the code for this chapter at `https://github.com/PacktPublishing/Data-Processing-with-Optimus`.

How data moves internally

Before we deep dive into how to load and save data, let's explore how external storage systems, databases, CPUs, RAM, GPUs, and local storage systems interact to get your data ready for processing.

We'll divide the process of loading data into four categories.

File to RAM

A file is loaded from disk to RAM using the CPU:

Figure 2.1 – Loading and saving data from a file using CPUs

File to GPU memory

A file is loaded from disk to GPU memory using the CPU. To save the file to disk, the data is sent via the GPU to the CPU, and then to disk:

Figure 2.2 – Loading and saving data from a file using GPUs

Database to RAM

The data is extracted from a database and sent via a JDBC driver to RAM on your local laptop or cluster:

Figure 2.3 – Loading and saving data from a database using CPUs

Database to GPU memory

In Optimus, the data is extracted from the database and sent via a JDBC driver to RAM using the CPU. To save the data back to the database, the data needs to be sent via the GPU to the CPU, and then to the JDBC driver to the database:

Figure 2.4 – Loading and saving data from a database using GPUs

There are new technologies that skip the CPU altogether, such as **GPUDirect Storage**. This can greatly improve performance when you're dealing with big data, though this has not been tested with Optimus at the time of writing. For more information, visit `https://developer.nvidia.com/blog/gpudirect-storage/`.

When using Optimus, we will try to do minimal data movement, because every move will incur time penalties. Later in this chapter, we will explain the implications this has for every engine.

Loading a file

As explained in the previous chapter, to load a dataframe, we have the `.load` accessor in the Optimus instance, which lets us to load from different sources, such as the local filesystem, databases, and alternative filesystems (S3, HDFS, and more).

The most useful and readable function from this accessor is `df.load.file`, which allows us to infer encoding data on the loading process so that we can forget about extra configuration when loading a new dataset. This following is an example:

```
from optimus import Optimus
op = optimus("pandas")
df = op.load.file("my_file.json")
```

In the preceding code, we are simply loading a JSON file. Internally, Optimus detects its format and creates the dataframe. It can also load formats such as CSV, JSON, XML, Excel, Parquet, Avro, ORC, and HDF5. If required, you can also call any specific method for its type, as shown here:

Method on **op.load**	File format	Type
csv	Comma-separated values files	Text
json	JavaScript Object Notation	Text
xml	Extensible Markup Language	Text
avro	Apache Avro	Binary
orc	Optimized Row Columnar (ORC)	Binary
excel	Excel files (xls, xlsx)	Binary
parquet	Apache Parquet	Binary
hdf5	Hierarchical Data Format	Binary

Figure 2.5 – Specific file format loading methods available in Optimus

When loading a JSON file, in our example, it does not infer much, only the encoding format and multi-line values support; but when a CSV file is loaded, it may need to infer some parameters, such as the separation character, the encoding format, line terminators, and other variants in the possible formatting. Finding the correct params can be very frustrating and time-consuming, but that is the way the `file` method was created.

The arguments of any of the methods available will be correctly inferred by `file` when loading the data, but what if you need to load them manually? The following table shows a list of the available arguments, including some that are not related to the format of the file:

Argument	Description
null_value	Default value: "None" Replaces all the matching values with null values.
n_rows	If set, limits the loading of a file to the number of rows to the passed value.
encoding	Default value: "UTF-8" Reads the file using the passed encoding. Only available for text formats (CSV, JSON, and XML).

Figure 2.6 – Arguments available for file loading in Optimus

If we have any trouble with a tedious file trying to infer its parameters via Optimus, we can always load using any of the specific methods available for every format. These methods have the same name as the format we're trying to load; one of the most frequently used formats is CSV, which may require a bunch of arguments to work properly in some cases.

Now, let's learn how to load data in multiple formats, just in case we need extra control. Let's start with **CSV** files.

Loading a local CSV file

A CSV file is a delimited text file that uses a comma to separate values. Each line of the file is a data record. Each record consists of one or more fields, separated by commas.

Let's explore the CSV format a little bit, because it has more parameters than any other loading function:

- You can load a fixed number of rows using n_rows:

```
from optimus import Optimus
op = optimus("dask")
df = op.load.csv("data/file.csv", n_rows=5)
```

- You can load without a header using the header argument:

```
df = op.load.csv("data/file.csv", header=None)
```

- You can load and assign which value will be assumed as null by passing it to null_vale:

```
df = op.load.csv("data/file.csv", null_value="Null")
```

- You can load using a specific separator by passing a string to sep:

```
df = op.load.csv("data/file.csv", sep=";")
```

We can also use wildcards. This is very handy if we need to select multiple files in a folder. Let's take a look.

Wildcards

Optimus also allows us to load multiple files with paths that match the wildcard that was passed to the first argument of the loading method, as follows:

```
from optimus import Optimus
Op = optimus("cudf")
df = op.load.csv("csv/*")
```

The preceding code will return a list of files – specifically, all the files in a folder called csv. More advanced wildcards can also be used if required:

```
df = op.load.csv("csv/file-*.csv")
```

The preceding code will load all the files that start with file- and have csv as their extension. This is especially useful, for example, for loading a series of files that are automatically named by another program.

Single-character wildcard

The other wildcard character that's supported is the question mark, ?. It matches any single character in that position in the name; for example:

```
df = op.load.csv("csv/file-?.csv")
```

In the preceding example, we're loading all the files whose name matches file-(anything).csv in the csv directory, where (anything) is a single character. This would load, for example, enumerated files such as dataset-1.csv or dataset-9.csv, but also load dataset-a.csv. However, it wouldn't load dataset-10.csv. For more specific formats, you can use character ranges.

Character ranges

When you need to match a specific character, use a character range instead of a question mark. For example, you can do this to find all the files that have a digit in the name before the extension:

```
df = op.load.csv("csv/file-*[0-9].*.csv")
```

In this case, we're loading the files that have been enumerated from zero to nine and include any strings before the extension. Files such as dataset-0.final.csv and dataset-1.wip.csv will be included.

Loading large files

When handling big files that are not using distributed dataframe technologies such as pandas, users will tend to use hacky ways to divide the file into multiple chunks, in order to load data in memory and process it. In Optimus, when you use the Dask or Vaex engines (which, as we learned in *Chapter 1*, *Hi Optimus!*, can handle out-of-core sources), you'll be able to load files that are bigger than your system's memory:

```
op = Optimus("dask")
df = op.load.csv("s3://my-storage/massive-file.csv")
```

In projects involving large datasets, you'll want to load from alternative filesystems, such as S3 or HDFS. This is possible by using connections that have already been set up.

Now, let's learn how to load a file from a remote data source.

Loading a file from a remote connection

If we need to access remote storage, a connection instance is needed. Using a connection instance, we can load several files from the same source without repeating every credential and parameter required to successfully connect to it:

```
from optimus import Optimus
Op = optimus("dask")
conn = op.connect.s3(endpoint_url="s3://my-storage/")
df = op.load.csv("files/foo_file.csv", conn=conn)
```

In the preceding example, we're simply creating a connection to an S3 bucket and then loading a file from it, passing it as a parameter called connection to load.csv. After that, we can reuse the conn instance to load another file from the same source:

```
df2 = op.load.file("files/file_bar.xml", conn=conn)
```

In the preceding line of code, we are loading a different file, this time using load.file, and passing the same conn instance. We learned how to configure a connection for remote data loading previously in this chapter.

Now that we have learned how to load data from local and remote files, let's learn how to load data from databases.

Loading data from a database

Another common use case when working with data wrangling is to load tables from a database. This function can be very handy because you do not need to save the data to a file, so it can be saved to memory. Therefore, Optimus allows us to load the most common database kinds in our workspace relatively easily:

```
from optimus import Optimus
Op = optimus("dask")
db = op.connect.mysql(host="localhost",
                      database="my_database")
df = op.load.database_table("foo_table", conn=db)
```

As you can see, we're creating a connection instance, but this time, it is connecting to a MySQL database in a given address. Then, we load a dataframe from a table on this database.

With Optimus, you can easily connect to more than 10 database technologies. In the following sections, we'll learn more about this.

Special dependencies for every technology

Every dataframe technology relies on different drivers to work, which means that dozens of additional files are needed to work out of the box. This will dramatically increase the size of the Optimus package. It is your responsibility to install the driver you need to connect to the database.

In this section, we will look at some of the databases that are supported by Optimus and where you can find the required file. Please take into account that URLs can change over time. If a link is broken, a simple Google search should get you the correct result.

Spark

When using Spark as your engine, you will need to pass the JDBC file path so that Spark knows how to handle a specific database.

Use the following command:

```
Optimus("spark", jars="path/to/jars/folder")
```

The following is a list of database technologies where you can find every driver. When working in a cluster, ensure that the file is accessible to every node:

- MySQL: https://dev.mysql.com/downloads/connector/j/
- MsSQL: https://search.maven.org/search?q=spark-mssql-connector
- PostgreSQL: https://jdbc.postgresql.org/
- Oracle: https://www.oracle.com/mx/database/technologies/appdev/jdbc-downloads.html
- SQLite: https://repo1.maven.org/maven2/org/xerial/sqlite-jdbc/
- Redshift: https://docs.aws.amazon.com/redshift/latest/mgmt/configure-jdbc-connection.html#download-jdbc-driver
- Cassandra: https://mvnrepository.com/artifact/com.datastax.spark/spark-cassandra-connector
- Presto: https://repo1.maven.org/maven2/com/facebook/presto/presto-jdbc/
- BigQuery: https://storage.googleapis.com/spark-lib/bigquery/spark-bigquery-latest.jar

- Impala: `https://www.cloudera.com/downloads/connectors/impala/jdbc/2-6-15.html`

> **Note**
>
> Some of these links may be outdated by the time you read this, so please identify the latest version and download its respective installer or `.jar` file.

pandas and Dask

pandas and Dask rely on **SQLAlchemy**, which uses a different library based on the database you want to access. The following is a guide to the libraries you will need, depending on the database you want to connect to, as well as the code repository, in case you need extra information:

Database	How to install	Repository
MySQL	python install pymysql	https://github.com/PyMySQL/PyMySQL
MSSQL	python install pymssql	https://github.com/pymssql/pymssql
Postgres	python install psycopg2	https://github.com/psycopg/psycopg2
Oracle	python -m pip install cx_Oracle --upgrade	https://oracle.github.io/python-cx_Oracle/
SQLite	Included in python 3.x	N/A
Redshift	pip install sqlalchemy-redshift	https://github.com/sqlalchemy-redshift/sqlalchemy-redshift
Cassandra	pip install cassandra-driver	https://github.com/datastax/python-driver
Presto	pip install presto-python-client	https://github.com/prestodb/presto-python-client
BigQuery	pip install pybigquery	https://github.com/mxmzdlv/pybigquery
Impala	pip install impyla	https://github.com/cloudera/impyla

Figure 2.7 – Database technologies for pandas and Dask

cuDF and Dask cuDF

In the case of cuDF/Dask-cuDF, Optimus use the same process as pandas/Dask. Here, the data needs to be loaded into RAM before it's loaded to the GPU memory. This means that the loaded dataframe must be serialized from pandas/Dask to be converted into cuDF/Dask-cudf, which will slow down the loading speed.

Vaex

Optimus also relies on SQLAlchemy to connect to databases, bring the data to pandas, and convert it into Vaex. Please follow the *pandas and Dask* section to install all the drivers you need.

Ibis

For PostgreSQL, SQLite, and MySQL (experimental as of January 2021), Ibis relies on SQLAlchemy, so please follow the instructions in the *pandas and Dask* section.

Now that we have explored the data loading options that are available, let's see how we can optimize an already loaded dataframe.

Memory usage optimization

Data types can make a big difference in memory usage. Let's take pandas as an example and see how many bytes some data types use per value:

- **int8 / uint8**: Consumes 1 byte of memory, with a range between -128/127 or 0/255.
- **bool**: Consumes 1 byte, true or false.
- **float16/int16/uint16**: Consumes 2 bytes of memory, with a range between -32,768 and 32,767 or 0/65,535.
- **float32/int32/uint32**: Consumes 4 bytes of memory, with a range between -2,147,483,648 and 2,147,483,647.
- **float64/int64/uint64**: Consumes 8 bytes of memory.

If the values of a column are from 0 to 255 and you are using float16 as the data type, you are using 2 bytes per value instead of 1. So, converting this column into **uint** can give you 50% more memory.

In this case, Optimus dataframes use the `optimize` method, which can help us save memory on our machine once our data has been loaded. Let's see how it works:

```
from optimus import Optimus
op = Optimus("dask")
df = op.create.dataframe({
    "a": [1000,2000,3000,4000,5000]*10000,
    "b": [1,2,3,4,5]*10000
})
df.size()
```

When we created the dataset, we entered `*10000` into the arrays of each column to increase the size of the dataframe. Later, we called `df.size()`, which outputs the number of bytes the dataframe occupies in memory, giving us the following result:

```
800128
```

So, if we run `optimize` and the columns meet the criteria, we can reduce that number. This is useful for dealing with several large dataframes:

```
df = df.optimize()
df.size()
```

This will result in the following output:

```
150128
```

As we can see, the memory usage was heavily reduced; when working with real data, this may vary a lot since data comes in unexpected forms. This function takes some time to complete, so it must be used carefully.

In the previous examples, we used `op.create.dataframe` to quickly create an example without files or external sources. Let's see how this works.

Creating a dataframe from scratch

If you want to create a dataframe from a Python dictionary, there's a method available on the `create` accessor called `dataframe` that requires a named dictionary that should contain an array of values for each entry:

```
df = op.create.dataframe({
"name":["OPTIMUS", "BUMBLEBEE", "EJECT"],
```

```
"function":["Leader", "Espionage",
            "Electronic Surveillance"]
})
```

In the preceding example, we're simply creating a new dataframe with two rows and three values each. This may be handy for quickly testing operations before loading a larger dataset.

Connecting to remote data sources

Unless your files are in your local disk, you are going to need to create a connection to your data sources in a storage service. Optimus provides a way to connect to these, even if it requires credentials.

Storage services are useful for making web-scale computing easier for developers, allowing them to store and retrieve massive data, at any time, from anywhere on the web.

To load from remote storage, we need to instantiate a connection. To achieve this, we must simply call, for example, `op.connect.hdfs` for connections to HDFS storage systems or `op.connect.s3` for connections to S3 storage systems. The resulting value can be saved for further reference when you're loading a file from remote storage. More alternatives will be addressed later in this section.

Let's look at an example:

```
conn = op.connect.s3(base_url="s3://bucket/", anon=False,
                     use_ssl=True)
```

In the preceding code, we are creating a connection to an Amazon S3 bucket by passing some storage options.

Next, we'll look at some of the available methods on `op.connect` and which filesystems it represents:

- `file`: Local or network filesystem (used by default when no connection has been configured)
- `s3`: Amazon S3
- `hdfs`: Hadoop File System
- `gcs`: Google Cloud Storage
- `adl`: Microsoft Azure (Data Lake Storage)

- `abfs`: Microsoft Azure (Blob Storage)
- `http` / `https`: HTTP(S) filesystem
- `ftp`: FTP filesystem

As we saw previously, we pass a `base_url` argument, including a piece of the URL. When you do this, you can pass a relative path when loading a file. This is useful if you want to change where your data comes from. Let's look at another example:

```
conn = op.connect.s3(base_url="s3://bucket/")
```

To use the connection from the previous example, we can use the following code:

```
df = op.load.file("files/my-remote-file.csv", conn=conn)
```

It will behave the same as the following code:

```
df = op.load.file("s3://bucket/files/my-remote-file.csv",
conn=conn)
```

In both examples, the same file will be loaded to `df`, thus inferring the format. Next, we'll learn how to benefit from the connections that are created using `op.connect`.

Connection credentials

You can connect to secure storage services using credentials, ensuring that only certain users can access your stored data.

> **Important Note**
>
> When you request your data, you may be exposing your credentials. If you're using this approach, be sure that the environment you are working in is totally secure. A more secure approach is to save the credentials in a local file in your cluster.

A more secure approach could be to place the credentials on a file in every node in the cluster. Every engine has its own unique way of handling the credentials. Please refer to each engine's documentation to learn about the exact steps you must follow to configure it.

Let's learn how to do this for each of the major remote storage providers and systems that are available.

Google Cloud Storage

Authentication for **Google Cloud Storage**, or **GCS**, requires us to use the `gcs` method. This requires a `token`, which can be provided using a `GCSFileSystem` from `gcsfs`. More details about this are available in the `gcsfs` documentation:

```
gcs = GCSFileSystem(...)
conn = op.connect.gcs(token=gcs.session.credentials)
```

Azure

Microsoft Azure Storage comprises Data Lake Storage (Gen 1) and Blob Storage (Gen 2). You can set them up using the `adl` and `abfs` methods, respectively.

Authentication for `adl` requires `tenant_id`, `client_id`, and `client_secret` to be in the `options` dictionary:

```
conn = op.connect.adl(tenant_id="<your tenant id>", client_
id="<your client id>", client_secret="<your client secret>")
```

Authentication for `abfs` requires `account_name` and `account_key` to be in `options`:

```
conn = op.connect.abfs(account_name="<your account name>",
account_key="<your account key>")
```

S3

For S3, the basic requirements are `key` and `secret`, which must be passed to `s3`:

```
conn = op.connect.s3(key="<your key>", secret="<your secret>")
```

However, if you don't want to write your URL every time you load a file, you can set the `base_url` argument:

```
conn = op.connect.s3(base_url="s3://<your-bucket-address>/",
key="<your key>", secret="<your secret>")
```

Alternatively, you can pass the bucket's name:

```
conn = op.connect.s3(bucket="<your-bucket-name>", key="<your
key>", secret="<your secret>")
```

Authentication for s3 requires various arguments. The following list shows every argument, along with an explanation, since not all of them are self-explanatory:

- anon: Whether access should be anonymous (default: False).

- key and secret: For user authentication.

- token: If authentication has been done with some other S3 client.

- use_ssl: Whether connections are encrypted and secure (default: True).

- client_kwargs: A dictionary is passed to the boto3 client, with keys such as region_name or endpoint_url.

- config_kwargs: A dictionary is passed to the s3fs.S3FileSystem, which passes it to the boto3 client's config option.

- requester_pays: Set to True if the authenticated user will assume transfer costs, which is required by some providers of bulk data.

- default_block_size, default_fill_cache: These concern the behavior of the buffer between successive reads.

- kwargs: Other parameters are passed to the boto3 Session object, such as profile_name, to pick one of the authentication sections from the configuration files referred to previously.

You can use any of the preceding terms like so:

```
conn = op.connect.s3(key="<your key>", secret="<your secret>",
anon=True, use_ssl=False)
```

In the preceding example, we are setting the basic authentication arguments for S3 and including anon and use_ssl.

HDFS

Authentication for hdfs requires host, port, and user to be in options:

```
conn = op.connect.hdfs(host="<your host>", port="<your port>",
user="<your user>")
```

You can also pass a path to a Kerberos ticket cache to kerb_ticket:

```
conn = op.connect.hdfs(host="<your host>", port="<your port>",
user="<your user>", kerb_ticket="<your kerb ticket path>")
```

In the preceding example, we are setting the basic arguments and including a Kerberos ticket cache on `kerb_ticket`.

As you can see, we can load from remote filesystems. Next, we will learn how to load datasets from databases using the most common engines that are available.

Connecting to databases

Connecting to databases can be very handy because you do not need to save the data to a file so that it can be loaded in memory. This might seem trivial if you are used to handling small files, but downloading big data datasets to files to be converted into specific formats can be a very tedious and boring task. This is where loading data directly from databases shines.

To create a connection to a database using Optimus, we can use any of the available methods for database handling. There are a few differences between this and remote filesystems, which we explained in the previous section:

```
from optimus import Optimus
Op = optimus("dask")
db = op.connect.postgres(address="localhost", user="root",
                         password="12345678",
                         database="mydatabase")
```

In the preceding example, we're connecting to a PostgreSQL database using certain credentials. If we want to list all the available tables in this database, we can simply call `db.tables()`, which will give us a list of every table in the connected database. It also allows us to test the connection to this database:

```
db.tables()
```

This gives us a result that's similar to the following:

```
['foo_table', 'bar_table']
```

Some of the available database engines Optimus can handle are MySQL, SQLite, PostgreSQL, Oracle, Microsoft SQL Server, BigQuery, Redshift, Cassandra, Presto, and Redis. Most of them have the same arguments available, but here's a list of the most frequently used options for creating a connection:

- `mysql`: MySQL
- `sqlite`: SQLite

- `microsoftsql`: PostgreSQL
- `postgres`: Oracle
- `oracle`: Microsoft SQL Server
- `bigquery`: BigQuery
- `redshift`: Redshift
- `cassandra`: Cassandra
- `presto`: Presto
- `redis`: Redis

As we mentioned previously, each method requires the following arguments in order to make a proper connection to a database:

- `host`: The host where the database is stored
- `port`: The port where the database is available
- `user`: Username
- `password`: Password
- `database`: The name of the database you want to connect to

Some of the engines that are available have specific arguments. Some of them are as follows:

- `schema`: Available on MySQL, Impala, BigQuery, PostgreSQL, and Redshift
- `tns`: Available on Oracle
- `service_name`: Available on Oracle
- `sid`: Available on Oracle
- `catalog`: Available on Presto
- `keyspace`: Available on Cassandra
- `table`: Available on Cassandra
- `project`: Available on BigQuery
- `dataset`: Available on BigQuery

Once the setup is complete, we can load a dataframe from whatever source it is stored on:

```
df = db.to_dataframe("my_table")
```

After loading the data, you can start processing it. Data processing will be addressed in *Chapter 3, Data Wrangling*.

Now that you know how to load and process your data, you need to save it. Let's see how this works.

Saving a dataframe

Saving on Optimus can be done by simply calling any of the methods available on the `save` accessor of a dataframe instance. In this section, we'll learn how to save to a local or remote filesystem, and also to a previously established database or remote storage connection.

When saving data in files, it is important to understand which format to use so that you can gain speed when reading or processing. There is plenty of information available about how to select the correct date format. I like the following for simplicity:

> *"Finding the right file format for your particular dataset can be tough. In general, if the data is wide, has a large number of attributes, and is write-heavy, then a row-based approach may be best. If the data is narrower, has a fewer number of attributes, and is read-heavy, then a column-based approach may be best."*

(Datanami, `https://www.datanami.com/2018/05/16/big-data-file-formats-demystified/`)

Saving to a local file

Saving methods are pretty similar to the loading methods available on the Optimus instance, but, as we discussed previously, saving methods are available in every dataframe instance, specifically in the `save` accessor:

```
df.save.csv("foo_output.csv", separator=";")
```

In the preceding code, we're saving the contents of the dataframe stored in `df` to a local file called `foo_output.csv`, in CSV format. This time, we're setting the separator to `;`. This allows us to save it in a different format from the incoming format.

When using a distributed engine such as **Dask** or **Spark**, it will save the data in multiple files in a folder. The number of files will depend on the number of partitions that have been configured. To save the dataframe in a single file, it's necessary to pass `True` to an argument called `single_file`, as shown here:

```
df.save.csv("foo_output.csv", single_file=True)
```

Now, let's learn how to save data to remote datastores.

Saving a file using a remote connection

Just as in the loading methods, there's a `conn` argument available for saving to a previously defined connection:

```
df.save.xml("files/foo_output.xml", conn=conn)
```

In the preceding example, we're saving our file in a filesystem that's been configured in the `conn` variable. This instance may contain a connection to a **HDFS** cluster, an **S3** bucket, or any similar system.

Now, let's see how Optimus can save data in a different format from the origin.

Saving in a different format

When saving, Optimus allows you to select a different file format from the source:

```
from optimus import Optimus
Op = optimus("dask")
df = op.load.csv("files/foo_input.csv")
```

To save the data in JSON format, you just simple use the `.save.json()` method:

```
df.save.json("files/foo_output.json")
```

In the preceding code, we're loading a CSV file and saving a JSON file. However, you can also change certain aspects of the file, such as the separator of a CSV file:

```
df = op.load.csv("files/foo_input.csv", sep=",")
df.save.csv("files/foo_output.csv", sep=";")
```

Alternatively, you can save the data to any other format that's supported in Optimus, such as Avro, Parquet, or OCR, using their respective methods.

Saving a dataframe to a database table

To save a dataframe as a table on a database that had previously been configured using `connect`, Optimus has a method called `database_table` in the `.save` accessor of the Optimus dataframe:

```
from optimus import Optimus
op = optimus("dask")
df = op.load.csv("files/foo_input.csv")
df.save.database_table("foo_output_table", db=db)
```

In the preceding code, we're simply saving our dataframe in a table called `foo_output_table`, in the database stored in `db`.

Loading and saving data in parallel

When loading and saving data, it can be very useful to use a distributed engine since it can parallelize the data loading process using multiple CPUs/GPUs cores at the same time, which means faster data loading times.

When creating an Optimus DataFrame, an object with a specific class is created, depending on the engine that was configured when you initialized the Optimus instance. For example, take a look at the following code:

```
op_dask = Optimus("dask", n_workers=4, threads_per_worker=1)
df = op_dask.load.csv("foo.csv")
op_pandas = Optimus("pandas")
df2 = op_pandas.load.csv("foo.csv")
```

Here, we're creating two different engines with two different dataframes: `df`, which is a **Dask Optimus DataFrame**, and `df2`, which is a **pandas Optimus DataFrame**. In this case, with Dask, you can configure the number of threads with `n_workers`, which will load your file using four parallel threads at the same time.

If your data fits in memory, it could be a good strategy to load it with the **Dask** engine. This allows you to load your data in parallel and then convert it into a pandas engine, as shown in the following code:

```
from optimus import Optimus
op_dask = Optimus("dask", n_workers=4, threads_per_worker=1)
df = op_dask.load.csv("foo.csv")
df = df.to_optimus_pandas()
```

> **Note**
>
> Dask can be configured to use thread or processes. Optimus configures Dask to use processes by default, allowing them to manage strings in parallel. Threads, on the other hand, are more convenient when processing numeric data. This is due to the Python **Global Interpreter Lock (GIL)**. Discussing how the GIL works is beyond the scope of this book.

When saving, you can parallelize the operation too, repartitioning the dataframe and saving it to disk. But first, you need to convert the pandas Optimus DataFrame into a Dask Optimus DataFrame. If you are working with a pandas dataframe, you could try this:

```
df = df.to_optimus_dask()
df = df.repartition(4)
df.save.csv("output.csv", single_file=False)
```

In the preceding example, we are creating four files in the `output.csv` folder. By default, Optimus's output is saved to a single file to ensure that you can easily port it to any other package for visualization or reporting.

Summary

Loading and saving are the most used operations when wrangling data. Optimus creates a flow that can assist in creating connections to data sources that can be reused for loading and saving data. Optimus also implements the most used file storage technologies such as Amazon S3 and Google Cloud Storage, and database connections such as **PostgreSQL** and **MySQL**, so that the user can have all the necessary tools at hand to make their work easier.

In terms of databases, we looked at the drivers that are required for every engine/database technology to save and load data from databases.

We also explored how to optimize dataframe memory usage – a very important step if you are handling big data since you could save as much as 50% of your memory space.

In the next chapter, we will start exploring some basic methods for filtering, deduplicating, and transforming data for further analysis.

Section 2: Optimus – Transform and Rollout

By the end of this section, you will have learned how to merge, wrangle, and transform data to prepare reports and create machine learning models.

This section comprises the following chapters:

- *Chapter 3, Data Wrangling*
- *Chapter 4, Data Combining, Reshaping, and Aggregating Data*
- *Chapter 5, Data Visualization and Profiling*
- *Chapter 6, String Clustering*
- *Chapter 7, Feature Engineering with Optimus*

3
Data Wrangling

Now that our data has been loaded in memory, we need to transform it so that it meets our needs. First, we will discuss transforming, searching, and replacing data, including strings, dates, emails, and URLs. When it comes to numerical data, we will review the mathematical and trigonometric functions available in Optimus.

To close this chapter, we will learn how to write some custom functions to expand the possibilities of data wrangling and improve Optimus.

The topics we will cover in this chapter are as follows:

- Exploring Optimus data types
- Operating columns
- Experimenting with user-defined functions

Technical requirements

Optimus can work with multiple backend technologies to process data, including GPUs. For GPUs, Optimus uses **RAPIDS**, which needs an NVIDIA card. For more information about the requirements for this, please go to the *GPU configuration* section of *Chapter 1, Hi Optimus!*.

You can find all the code for this chapter at https://github.com/PacktPublishing/Data-Processing-with-Optimus.

Exploring Optimus data types

Data types are the soul of a dataframe: they define how a value is represented in memory and, more importantly, how much memory it will use. Every dataframe technology supported in Optimus has different data types aimed to represent specific data. The most common are numeric values, string values, and datetime values. You can find which data types are supported in each technology by going to its respective website or documentation. This information can be found in the *Further reading* section of this chapter.

Besides internal data representation, Optimus tries to enrich the data to give the user a better overview of how it can be wrangled. For example, when you see a column that's of the *email* type, internally, it is just a string column, but when the profiled is requested, it gives us feedback about how many mismatches (data points that do not match the type) are on a column. We'll talk more about the profiler later in this book.

Converting data types

Optimus is designed to let you convert into any data type you require, regardless of what form the data takes. In the case of most dataframe technologies, it will raise an error if you try to convert from non-compatible data types, for example, from a non-integer value such as `"optimus"`, which is a string, to an integer.

Let's start creating some data for example purposes using a dictionary. In this data, we'll be including some of the autobots and their jobs, but also including some placeholder numeric values and a birth column to test some functions:

```
df = op.create.dataframe({
    "name": ["optimus", "bumblebee", "eject", 2],
    "job": ["Leader", "Espionage", 1, 3],
    "id": [1, 2, 3, 4],
    "birth": ["22/10/10", "22/08/08", "23/07/07", "22/10/10"]
})
```

| name | job | id | birth |
(object)	(object)	(int64)	(object)
optimus	Leader	1	22/10/10
bumblebee	Espionage	2	22/08/08
eject	1	3	23/07/10
2	3	4	22/10/10

Let's make a little test. By using df.data, you have access to the dataframe's internal representation. If you are using the pandas engine, you will get a pandas dataframe:

```
print(df.data)
print(type(df.data))
          name        job   id     birth
0      optimus     Leader    1  22/10/10
1    bumblebee  Espionage    2  22/08/08
2        eject               1    3  23/07/07
3            2               3    4  22/10/10
<class 'pandas.core.frame.DataFrame'>
```

As we can see, we printed the internal dataframe and its type, which gives us a different result than calling print(df).

Let's see what happens when we try to convert the values of a string column of that internal dataframe into integer values:

```
df.data["name"].astype(int)
```

This will raise the following error:

```
ValueError: invalid literal for int() with base 10: 'optimus'
```

With pandas, you can use pd.to_numeric() to handle this type of scenario. However, it feels like it breaks the flow of how we operate over the dataframe.

Now, let's see how Optimus can handle this.

Let's convert data type of the name column of that dataset into an integer:

```
df.cols.to_integer("name")
```

We'll see a result similar to the following:

```
   name  job                 id  birth
(int64)  (object)       (int64)  (object)
-------- ----------     -------- ----------
      0  Leader               1  22/10/10
      0  Espionage            2  22/08/08
      0  1                    3  23/07/10
      2  3                    4  22/10/10
```

The `to_integer` method converts non-integer values into 0.

Now, let's convert data type of the "name" column into float using `to_float`:

```
df.cols.to_float("name")
     name   job               id   birth
  (float64)  (object)       (int64)  (object)
  ----------  ----------   ----------  ----------
       nan  Leader              1   22/10/10
       nan  Espionage           2   22/08/08
       nan  1                   3   23/07/10
         2  3                   4   22/10/10
```

Here, non-float compatible values are converted into nan (Not a Number). If the dataframe doesn't internally support nan values (for example, a pandas dataframe when using pandas before version 1.0), non-integer values will be converted into 0.

On the other hand, converting an int column into a string will probably not change how the value is shown to the user, but it will change the internal column's data type:

```
df.cols.to_string("id")
```

As we will see, the id column data type changes from int to object:

```
name         job               id   birth
(object)     (object)       (object)  (object)
----------   ----------     ----------  ----------
optimus      Leader              1   22/10/10
bumblebee    Espionage           2   22/08/08
eject        1                   3   23/07/10
2            3                   4   22/10/10
```

For datetime, we can convert a string into a datetime object so that we can apply this function to extract dates elements such as day, month, year, hours, minutes, and seconds.

For this function, you must include the format in the second argument. You can also use the `format` keyword for this:

```
df.cols.to_datetime("birth", "YY/mm/dd")
```

As we will see, this will convert the `birth` column into a datetime data type:

```
name           job              id   birth
(object)       (object)         (int64)   (datetime64[ns])

----------     ----------     ----------   --------------------
optimus        Leader            1   2022-10-10 00:00:00
bumblebee      Espionage         2   2022-08-08 00:00:00
eject          1                 3   2023-07-10 00:00:00
2              3                 4   2022-10-10 00:00:00
```

When we convert the data types of a dataframe, we may be implicitly transforming the values of a column, but that's not the intended behavior. If we want to clean or transform the columns of a dataframe, we can use some operations like the ones we'll be exploring next.

Operating columns

To understand column operations, first, we'll look at some of the most common operations, which are for selecting columns. In Optimus, you have the powerful `select` function, which provides multiple options for managing most selection cases.

Selecting columns

The simplest use case for selecting a column is using its name:

```
df.cols.select("job").print()
```

You will get the job column data in response:

```
job
(object)

----------
Leader
Espionage
1
3
```

If you want to select multiple columns by their whole names, you can use a Python list, like so:

```
df.cols.select(["name", "job"]).print()
```

This will return the name and job column data:

```
name          job
(object)      (object)
----------    ----------
optimus       Leader
bumblebee     Espionage
eject         1
2             3
```

As we already know, if we want to save the modified dataframe, we must assign the result of the operation back to df, as shown here:

```
df = df.cols.select("name", "job", "id")
```

Another handy parameter is invert, which you can use to invert the selection for a single or a list of files:

```
df.cols.select("job", invert=True).print()
```

This will return all the columns except the name column:

```
name              id
(object)          (int64)
----------        ----------
optimus           1
bumblebee         2
eject             3
2                 4
```

You can also apply regular expressions to select columns. For this example, with the ^n. regular expression, you can get all the columns that start with the letter n:

```
df.cols.select(regex="^n.").print()


name
(object)
----------
optimus
bumblebee
```

```
eject
2
```

To finish, you can select columns by data type, like so:

```
df.cols.select(data_type="int").print()
```

With this, you can get all the columns of the integer data type, in case you want to filter out non-numeric data. However, this may not be useful in a reports dataframe, for example:

```
        id
     (int64)
  ---------
         1
         2
         3
         4
```

Next, we'll look at moving columns.

Moving columns

To change the order of the columns of a dataset, you can use df.cols.move, which can help us easily organize our columns. To use it, you must input the columns (or a single column) into the function, telling it where to place the selected columns relative to the reference column. Let's look at an example:

```
df.cols.move(["name", "job"], "after", "id")
```

This will organize the columns of the dataset into the form ["id", "name", "job"]. This order will also work if you pass "end" instead of "after" to the position argument (the second one), and do not pass anything to ref_col (the third one):

```
df.cols.move(["name", "job"], "end")
```

The position argument has four possible values:

- "before": This moves the column or columns that were passed so that they're before the one in ref_col.

- "after": This moves the column or columns that were passed so that they're after the one in ref_col.

- "beginning": This puts the column or columns that were passed to the beginning. Doesn't require any value for ref_col.

- "end": This puts the column or columns that were passed to the end. Doesn't require any value for ref_col.

Now, let's look at renaming columns.

Renaming columns

You can easily rename a column or a list of columns by simply using df.cols.rename. This function uses just two arguments, in which you must enter the name of the column whose name you want to change and the name you want it to have. It also supports arrays for both arguments, allowing multiple columns to be renamed by calling the function only once. Let's look at both examples:

```
df = df.cols.rename(["name", "job"], ["string_name", "string_
job"])
df = df.cols.move("id", "int_id")
```

The resulting dataframe that's been saved on df will contain the ["int_id", "string_name", and "string_job"] columns.

Removing columns

Optimus has two main functions that allow us to filter the columns in a dataset: drop and select. The first one allows us to choose which columns will be removed from the dataset, while the second one will do the opposite, removing all the columns that are not input.

If we call df.cols.drop("job"), the "job" column will be removed from the dataset, leaving "id" and "name". If, instead, we call df.cols.select("job"), only that column will be in the resulting dataset. Also, multiple columns on an array are supported for both functions, as shown here:

```
df.cols.select(["job", "name"])
```

In the preceding example, we are implicitly removing "id" from the dataset by passing "id" and "name" to df.cols.select.

Input and output columns

One handy and powerful function you can find in Optimus is that you can operate over one, multiple, or the whole set of columns and easily control the output.

Some functions (specifically, every function that mutates the column contents) can be configured so that it can be applied to a list of columns. Optimus also allows us to save those mutated columns in a set of new columns that have been appended to the dataset.

For example, `df = df.cols.upper("name")` will mutate the name column and overwrite it, transforming the strings into uppercase (later, we will learn how `upper` and other string functions work), and then save the resulting dataframe to `df`, overwriting the previous one. Let's see every case, including dynamically generated output column names, and entering the entire dataset into one operation.

You can **modify a single column** by running the following command:

```
df = df.cols.upper("name")
```

Here, we are modifying `"name"` and creating 0 new columns.

You can **modify multiple columns** by running the following command:

```
df = df.cols.upper(["name", "job"])
```

Here, we are modifying `"name"` and `"job"` and creating 0 new columns.

You can **modify a single column and save it to a new one** by running the following command:

```
df = df.cols.upper("name", output_cols="name_upper")
```

We are not modifying an existing column here; instead, we're creating `"name_upper"` from the result of operating `"name"`:

You can **modify multiple columns and save to a new set of columns** by running the following command:

```
df = df.cols.upper(["name", "job"], output_cols=["name_upper",
"job_upper"])
```

We are not modifying a previously existing column here; instead, we're creating `"name_upper"` and `"job_upper"` from the result of operating `"name"` and `"job"`.

You can **modify multiple columns and save to a new set of columns using a suffix** by running the following command:

```
df = df.cols.upper(["name", "job"], "upper")
```

Again, we are not modifying an existing column; instead, we're creating `"name_upper"` and `"job_upper"`, but this time, the names are generated dynamically, from the result of operating `"name"` and `"job"`.

You can **modify every column** without adding any new columns by using the star symbol; that is, `"*"`:

```
df = df.cols.upper("*")
```

Here, we are modifying every column in the dataset.

Note that you can also **chain operations**, as shown here:

```
df = df.cols.upper("name").cols.lower("job")
```

This specific command will convert the name of the column into uppercase and the job column into lowercase.

Now that we have a good understanding of how data types work and how to manage input and output columns, let's see which functions we can apply to them.

Managing functions

If you want to create, duplicate, delete, or rename one or more columns in your dataset, the syntax is pretty similar to the basic operations shown previously.

Calling the `drop` method will delete columns from the dataset:

```
df = df.cols.drop("job")
```

This will only delete the column called `"job"`.

If you want to delete every column except one (or more), you can call `keep`:

```
df = df.cols.keep("id")
```

In the preceding example, we're deleting every column except `id`.

To rename columns, the `output_cols` argument is not optional. Let's look at an example:

```
df = df.cols.rename("name", "first_name")
```

In the preceding code, we're simply renaming the name column to `first_name`, but if we wanted to rename a series of columns instead of just one, we would simply pass two lists of names.

The `duplicate` method works in a similar fashion – it just requires the `output_cols` argument. Here you can see an example with two columns instead of one:

```
df = df.cols.duplicate(["name", "id"], ["first_name", "id_
number"])
```

For creating columns, there are plenty of options. The most common is `set`, which can also set the content of an existing column, but in this case, we are using it to create a new column filled with None:

```
df = df.cols.set("last_name", None)
```

If you created an empty column, you can replace all the None values with any other value by using `fill_na`:

```
df = df.cols.fill_na("last_name", "Placeholder Value")
```

The `fill_na` method is also useful for an already existing and partially filled column, as it will only set the values of the empty rows in the column.

String functions

The string function is self-descriptive. Let's see all the functions and some brief descriptions of each. We'll look at examples of just a few of them since some operations are pretty similar:

Function	Description
upper	Convert the data to upper case
lower	Convert the data to lower case
title	Convert the data to title case
proper	Convert the data to proper case
capitalize	Capitalizes every value of the column
pad	Pad the data
replace	Replace the data string-wise
nest	Join multiple columns in one
unnest	Split a column in several columns by a separator

Figure 3.1 – String functions

Let's take the upper function to illustrate how this operation works. If you want to operate over one column, do the following:

```
print(df.cols.upper("name"))
```

This transforms the column with "name" as its title:

name(object)	job(object)	id(int64)
-------------	-------------	-----------
OPTIMUS	leader	1
BUMBLEBEE	espionage	2
EJECT	1	3
2	3	4

This was the most basic operation. Let's see how we can create a new column with the output of the function instead of replacing the original:

```
print(df.cols.upper("name", "upper_name"))
```

This creates a new column named `"upper_name"` that contains the values from `name` in uppercase:

name (object) id (int64)	upper_name (object)	job (object)	
optimus 1	OPTIMUS	leader	
bumblebee 2	BUMBLEBEE	espionage	
eject 3	EJECT	1	
2 4	2	3	

All the string functions can be applied to any data type. For example, you can call `df.cols.pad` to an integer column, and Optimus will transform the column's data type to string before applying the `pad` function:

```
print(df.cols.pad("id", 3, "0"))
```

This will modify the existing `id` column:

name (object) id (object)	upper_name (object)	job (object)	
optimus 001	OPTIMUS	leader	
bumblebee 002	BUMBLEBEE	espionage	
eject 003	EJECT	1	
2 004	2	3	

As we can see, Optimus tries to be less strict and convert the data type instead of raising an error for operating columns that could be incompatible. This is useful for concatenating the string values of a column with the numeric values of another; for this, we can use the `nest` function. Let's take a look.

Merging and splitting columns

When working with dates, names, addresses, or any fairly complex string data, merging and splitting are useful functions. Optimus provides two functions for this: `df.cols.nest` and `df.cols.unnest`.

To illustrate this, first, let's look at an example dataset with two columns that will be split and merged for example purposes – one with a pair of numbers in string format, which represents a point in a two-dimensional space, and another one representing the point in another dimension. This data is purely to demonstrate how to merge and split columns:

```
df = op.create.dataframe({
    "xy_position": ["42.8, 7.7", "23.3, 25.1", "35.6, 50.5",
"52.7, 67.4"],
    "depth": [10.5, 50.1, 20.2, 97.0]
})
df.print()
xy_position              depth
(object)            (float64)
-------------       -----------
42.8, 7.7                10.5
23.3, 25.1               50.1
35.6, 50.5               20.2
52.7, 67.4               97
```

First, let's see how we can separate all the dimensions into three columns by applying the `unnest` function to the first one:

```
df.cols.unnest("xy_position", separator=", ", output_cols=["x",
"y"], drop=True).print()
```

We'll get the following result:

x	y	depth
(object)	(object)	(float64)
----------	----------	-----------

42.8	7.7	10.5
23.3	25.1	50.1
35.6	50.5	20.2
52.7	67.4	97

As we can see, the resulting columns can now be operated as numeric columns since they have matching formats.

On the other hand, let's say we run the `nest` function to join the two original columns:

```
df.cols.nest(["xy_position", "depth"], separator=", ", output_
col="xyz_position", drop=True).print()
```

We will get the following result:

```
xyz_position
(object)

----------------
42.8, 7.7, 10.5
23.3, 25.1, 50.1
35.6, 50.5, 20.2
52.7, 67.4, 97.0
```

This results in a dataframe with just one column that includes all three dimensions.

When you split any column, it transforms its values into a string before creating new columns, so it behaves as expected. Similarly, when joining two or more columns, the resulting column will tend to be a string, regardless of the type of the input columns.

Search and replace

When we have a lot of string data, it may be useful to replace characters or words inside the values of the columns. For this, we can use `df.cols.replace`, which allows us to search for one or more values, words, or substrings, and replace them with a given string.

Let's look at an example with some arbitrary names separated by characters that will be replaced later. We'll use some autobots again for this:

```
df = op.create.dataframe({
    "name": ["Optimus, Prime", "Arcee, Ariel", "Bumblebee/
Maggiolino"]
})
df.print()
```

This will print out the following:

```
name
(object)
----------------------
Optimus, Prime
Arcee, Ariel
Bumblebee/Maggiolino
```

As we can see, the names in the column were entered using two different separators, that is, ", " and "/". Now, let's run the replace function:

```
df.cols.replace("name", [", ", "/"], " ", search_by="chars").
print()
```

Here, we get the correct format for this specific case:

```
name
(object)
--------------------
Optimus Prime
Arcee Ariel
Bumblebee Maggiolino
```

In the previous example, we replaced all the ", " and "/" matches with a single white space character (" ").

Numeric functions

Numeric functions can be divided into trigonometric and mathematical functions. Like the string operation, it supports all the column types but depends on their contents to work as expected:

Function	Description
sin	Calculate the sine of all values in one or more columns
cos	Calculate the cosine of all values in one or more columns
tan	Calculate the tan of all values in one or more columns
asin	Calculate the arcsine of all values in one or more columns
acos	Calculate the arccosine of all values in one or more columns
atan	Calculate the arctangent of all values in one or more columns
sinh	Calculate the hyperbolic sine of all values in one or more columns
cosh	Calculate the hyperbolic cosine of all values in one or more columns
tanh	Calculate the hyperbolic tangent of all values in one or more columns
asinh	Calculate the hyperbolic arcsine of all values in one or more columns
acosh	Calculate the hyperbolic arccosine of all values in one or more columns
atanh	Calculate the hyperbolic arctangent of all values in one or more columns

Figure 3.2 – Trigonometric functions

Before illustrating how the sin function works, let's create some example arbitrary data with numbers and a string. We will try to operate this later:

```
df = op.create.dataframe({ "values": [0.5, 2, "3.14"] })

df.print()

    values
   (object)
  ----------
```

```
    0.5

    2

    3.14
```

As we can see, the `values` column is being inferred as an object column because we entered `"3.14"` as a string, but this shouldn't be a problem when we apply the `sin` function:

```
df.cols.sin("values").print()
```

We'll get the following output:

```
      values
    (float64)

    ----------

  0.479426

  0.909297

  0.00159265
```

In the preceding example, we can see how Optimus can infer a numeric type when we apply a numeric function.

There are other functions available besides the trigonometric ones:

Function	Description
sqrt	Calculate the square root of all values in one or more columns
mod	
ln	Calculate the natural logarithm of all values in one or more columns
log	Calculate the logarithm of all values in one or more columns, uses an extra argument for the base (default 10)
round	
ceil	Rounds up all the values in one or more columns
floor	Rounds down all the values in one or more columns
exp	Calculates a number raised to the nth power, where n is every value of one or more columns

Figure 3.3 – Mathematical functions

Let's see how `log` works since it does have an extra argument, unlike some of the others:

```
df.cols.log("values", 5).print()
```

We'll get the same columns, but this time, all the numbers in `values` will have their logarithms:

```
     values
   (float64)
 ----------
  -0.430677
   0.430677
   0.710946
```

As we can see, all the original values were replaced by its logarithm when 5 was used as a base.

Date and time functions

There are also several functions related to date and time:

Function	Description
year	Extracts the year from the date values in one or more columns
month	Extracts the month from the date values in one or more columns
day	Extracts the day from the date values in one or more columns
hour	Extracts the hour from the date values in one or more columns
minute	Extracts the minute from the date values in one or more columns
second	Extracts the minute from the date values in one or more columns
weekday	Extracts the minute from the date values in one or more columns
date_format	Changes the date format of one column to another represented in the `output_format` argument. You can also force the input columns date format using `current_format`

Figure 3.4 – Date and time functions

Using any of these functions will facilitate changing the format or extracting data from the values of a column. For example, let's say we have the following dataset, which contains some arbitrary dates:

date (object)
01/21/2021
09/09/2020
03/07/2020

We can call `df.cols.year` to extract only the year from the values stored in the date column:

```
df.cols.year("date")
```

We'll will get the following result:

date (object)
2021
2020
2020

On the other hand, `date_format` is a bit more advanced, requiring us to input a new format, as shown here:

```
df.cols.date_format("date", output_format="%m-%d-%Y")
```

We'll get the following result:

date (object)
01-21-2021
09-09-2020
03-07-2020

Next, we'll look at URL functions.

URL functions

In Optimus, URLs are handled internally as strings. However, you can use these functions to extract data from URLs if they've been formatted correctly:

Function	Description
url_scheme	Extracts the schema from the URL values in one or more columns
sub_domain	Extracts the subdomain from the URL values in one or more columns
domain	Extracts the domain from the URL values in one or more columns
port	Extracts the port from the URL values in one or more columns
url_path	Extracts the URL path in one or more columns
url_params	Extracts the URL param from one or more columns

Figure 3.5 – URL functions

For example, you could create a new column with the domain, subdomain, and port of a URL. Let's look at an example with a URL value:

```
from optimus import Optimus
op = Optimus("pandas")


data = {"A": ["https://www.hi-optimus.com:8080/index.php?a=1"]
}
df = op.create.dataframe(data)

df["port"] = df.cols.port("A")
df["subdomain"] = df.cols.subdomain("A")
df["domain"] = df.cols.domain("A")
```

Next, we'll look at email functions.

Email functions

As well as URLs, Optimus can handle emails, and provides some functions to facilitate data extraction:

Function	Description
email_username	Extracts the schema from the email values in one or more columns
email_domain	Extracts the domain from the email values in one or more columns

Figure 3.6 – Email functions

Let's look at an example of using some arbitrary emails:

```
df = op.create.dataframe({"A": ["optimus@cybertron.com"]})
```

```
df["username"] = df.cols.email_username("A")
df["domain"] = df.cols.email_domain("A")
```

In the previous examples, we followed a different approach; that is, creating a new dataset by setting it as a property. If the output of the right-hand side of the assignment is a dataframe with a single column, it will join the two datasets.

Now that we've looked at all the functions that are available in Optimus, let's learn how to write our very own functions.

Experimenting with user-defined functions

Optimus tries to provide the most commonly used functions out of the box so that you can focus on your work instead of writing code. Of course, there are times when you will need to write custom functions to accomplish a task.

Before we deep dive into **user-defined functions** (UDF), let's explore a couple of scenarios regarding how data can be processed. Two such scenarios are known as vectorized and non-vectorized execution. This is important to understand because it can have a very big impact on performance.

Vectorized execution refers to operations that are performed on multiple components of a vector at the same time, in one statement. A vector is just a list of elements like the following:

```
[0, 1, 2, 3, 4, 5]
```

In the case of **non-vectorized** operations, the functions are executed in every element, one at a time. In the previous list, we need to pass every element to execute an operation. That's why using vectorized functions can improve processing times in our workflow. Let's see how this works in Optimus.

In Optimus, a vector is a column like the following, with a sequence of numbers. We'll use this for example purposes:

```
data = {"A": [0, 1,2,3,4,5], "B":[6,7,8,9,10,11] }df =
op.create.dataframe(data)
df.print()
```

```
        A           B
    (int64)     (int64)
    ---------   ---------
        0           6
        1           7
        2           8
        3           9
        4          10
        5          11
```

Now we will learn about using apply in Optimus.

Using apply

To apply a UDF in Optimus, you can define your own function and use the `df.cols.apply` method. For example, to define a function that just adds 2 to our whole column, run the following code:

```
def add_two(pandas_series):
    print(type(pandas_series))
    return pandas_series + 2


df.cols.apply("A", add_two).print()
```

This will output our whole dataset with the values of the "A" column increased by 2:

A (int64)	B (int64)
---------	---------
2	6
3	7
4	8
5	9
6	10
7	11

When you write your own UDF, you must take care of how to handle the data that's passed to the function. In this case, because we are using the pandas engine, the value that's passed to the function is a pandas series that supports the plus operator being executed in a vectorized way.

But there are functions that can't be vectorized. For this, we can use the mode="map" parameter to execute a function over every element:

```python
def add_two(single_value):
    print(type(single_value))
    return single_value + 2


df.cols.apply("A", add_two, mode="map").print()
```

In this case, print will be executed five times – once for every element:

```
<class 'int'>
<class 'int'>
<class 'int'>
<class 'int'>
<class 'int'>
<class 'int'>
```

A (int64)	B (int64)
---------	---------
2	6
3	7

4	8
5	9
6	10
7	11

Also, remember that you can use the star, `"*"`, to apply the function to every column in the dataset:

```
def add_two(value):
    return value + 2


print(df.cols.apply("*", add_two))
```

This will add 2 to every element in every column (A and B) and print it:

A (int64)	B (int64)
---------	---------
2	8
3	9
4	10
5	11
6	12
7	13

With that, we've learned how to process data in one or multiple columns. Now, let's learn how to use multiple values from different columns to make a calculation.

Calculations over multiple columns

Let's explore a popular function since we're talking about vectorized functions: the **Haversian distance**. This represents the distance between two points on the surface of a sphere and can be used to calculate the distance between two points on Earth.

The interesting thing about this function is that it uses multiple calculations that can be vectorized.

Here, we will use a popular data source from hotels in New York, and we are going to calculate the distance from a particularly popular restaurant to every hotel in the data.

Let's start by loading a dataframe from a file:

```
df = op.load.file("DCIGNP2AYL.txt")
```

Now, we can import the functions from Optimus and define our custom function:

```
from optimus.functions import F

def haversine(lat1, lon1, lat2, lon2):
    MILES = 3959
    lat1, lon1, lat2, lon2 = map(F.radians, [lat1, lon1, lat2,
lon2])
    dlat = lat2 - lat1
    dlon = lon2 - lon1
    a = F.sin(dlat/2)**2 + F.cos(lat1) * F.cos(lat2) *
F.sin(dlon/2)**2
    c = 2 * F.asin(F.sqrt(a))
    total_miles = MILES * c
    return total_miles
df["distance"] = haversine(40.671, -73.985, df["latitude"],
df["longitude"])
```

Then, we can use `select` and `print` to get the column with the previously calculated distances:

```
print(df.cols.select("distance"))
```

This will print the following:

```
    distance
    (float64)
-----------
139.607
139.747
142.191
137.276
...
```

Exploring the whole script, we can see the use of `from optimus.functions import F`. This will give you access to all the functions we explored in this chapter, all of which can be used to create custom and more complex calculations.

Supporting multiple engines

It is important to note that at the time of writing, in January 2021, **cuDF** and **Dask cuDF** do not support UDF functions over strings. Also, UDF numeric functions are not supported, so we need to rely on cuDF directly. Let's take a look:

```python
import numpy as np
import cudf
from cudf.datasets import randomdata


df = randomdata(nrows=10, dtypes={"a": float, "b": bool, "c": str}, seed=12)

def udf(x):
    if x > 0:
        return x + 5
    else:
        return x - 5

df["a"].applymap(udf)
```

For more complex logic (such as accessing values from multiple input columns or rows), you'll need to use a Numba JITed CUDA kernel:

```python
from numba import cuda

@cuda.jit
def multiply(in_col, out_col, multiplier):
    i = cuda.grid(1)
    if i < in_col.size: # boundary guard
        out_col[i] = in_col[i] * multiplier
```

The `create` function will multiply every element in column a and output the result to column e.

Now, to apply the function, we will use the `forall` method. With `forall`, you need to specify the size of the output and pass the data you want to process into the dataframe:

```
size = len(df['a'])
df['e'] = 0.0
multiply.forall(size)(df['a'].data, df['e'].data, 10.0)
```

As you can see, this is a very different approach to applying a UDF. We hope that, in the future, we can abstract it and make it more user-friendly.

For more information on how to apply UDF to cuDF and Dask-cuDF, please read the RAPIDS docs: `https://docs.rapids.ai/api/cudf/stable/guide-to-udfs.html`.

Summary

In this chapter, we learned about basic Optimus functions that were designed to cover the most common work in dataframes, such as selecting, moving, and dropping columns, and applying functions over strings, numbers, dates, and more specific data, such as URLs and emails.

We also learned how to write custom functions and how to use vectorized functions to access the full potential of our hardware.

In the next chapter, we will learn how to join multiple datasets so that we can shape our data to our needs.

Further reading

Every engine handles a data type internally to represent numbers, strings, and dates. The following links can help you find out about the different data types in every engine:

- **pandas/Dask**: `https://pbpython.com/pandas_dtypes.html`.
- **CuDF/Dask-cuDF**. There is an open issue at `https://github.com/rapidsai/cudf/issues/3360`.
- **Spark**: `https://spark.apache.org/docs/latest/sql-ref-datatypes.html`.

4
Combining, Reshaping, and Aggregating Data

When we must deal with multiple datasets simultaneously, it's important to have the right tools that allow us to combine said datasets into a homogeneous and uniform one. As we saw in the previous chapters, Optimus provides us with transformation operations that allow us to prepare a dataset whose format does not coincide with another, so that we can combine them correctly later. Once transformed, it is possible to combine them in various ways, such as via concatenation or union.

In this chapter, we'll learn how to concatenate and merge multiple datasets using Optimus and review more complex transformations such as reshaping and pivoting. To finish, we will learn how to aggregate data and how to apply aggregation over a specific group of data.

Some of these concepts are maybe already known to those of you who have come from the relational database world. If you are a novice, then don't worry – we will try to explain this concept using some graphical support to get you on track as quickly and easily as possible.

The topics we will be covering in this chapter are as follows:

- Concatenating data
- Joining data
- Reshaping and pivoting
- Aggregating and grouping

Technical requirements

Optimus can work with multiple backend technologies to process data, including GPUs. For GPUs, Optimus uses **RAPIDS**, which needs an NVIDIA card. For more information about the requirements for this, please go to the *GPU configuration* section of *Chapter 1, Hi Optimus!*.

You can find all the code for this chapter at `https://github.com/PacktPublishing/Data-Processing-with-Optimus`.

Concatenating data

We call concatenation the process of joining two datasets, whether we're taking their columns and including them in another or combining their rows, resulting in a dataset whose number of rows is equal to the sum of the number of rows of the datasets to be concatenated.

Let's look at an example of row concatenation:

```
df_a = op.create.dataframe({
    "id": [143, 225, 545],
    "name": ["Alice", "Bob", "Charlie"],
    "city": ["Plymouth", "Bradford", "Norwich"]
})
df_b = op.create.dataframe({
    "id": [765, 329, 152],
    "name": ["Dan", "Erin", "Frank"],
    "city": ["Bath", "Manchester", "Ripon"]
})
```

df_a and df_b can be concatenated as follows:

```
df_a.rows.append(df_b).print()
```

This results in the following output:

```
id          name        city
(int64)     (object)    (object)
---------   ----------  ----------
143         Alice       Plymouth
225         Bob         Bradford
545         Charlie     Norwich
765         Dan         Bath
329         Erin        Manchester
152         Frank       Ripon
```

In the previous example, we can see how the datasets are combined, giving rise to a third dataframe with a larger number of rows but the same number of columns. This is because the columns of the input datasets coincide.

But what if we omit a column from df_b?

```
df_b = df_b.cols.drop("city")
df_a.rows.append(df_b).print()
```

We will get the following result:

```
id          name        city
(int64)     (object)    (object)
---------   ----------  ----------
143         Alice       Plymouth
225         Bob         Bradford
545         Charlie     Norwich
765         Dan         nan
329         Erin        nan
152         Frank       nan
```

As you can see, all the missing values are filled with nan. A similar result can occur if the datasets have different column names:

```
df_b = df_b.cols.rename("city", "city_b")
df_a.rows.append(df_b).print()
```

We'll get the following result:

```
id          name        city        city_b
(int64)     (object)    (object)    (object)

---------   ----------  ----------  ----------

143         Alice       Plymouth    nan
225         Bob         Bradford    nan
545         Charlie     Norwich     nan
765         Dan         nan         Bath
329         Erin        nan         Manchester
152         Frank       nan         Ripon
```

Now that we know the basics of how to concatenate two dataframes with the same column names, we'll learn how to map the columns of two different dataframes so that we can concatenate them without renaming either.

Mapping

When we have two equal datasets with different column names, we can pass a second argument called map, which allows us to map the columns of each input dataframe to a set of output columns.

Let's look at an example:

```
df_a = op.create.dataframe({
    "id": [143, 225, 545],
    "name": ["Alice", "Bob", "Charlie"],
    "city": ["Plymouth", "Bradford", "Norwich"]
})
df_b = op.create.dataframe({
    "id_number": [765, 329, 152],
    "name": ["Dan", "Erin", "Frank"],
    "title": ["Bath", "Manchester", "Ripon"]
})
```

df_a and df_b can be concatenated as follows:

```
names_map = {
    "id_number": ("id", "id_number"),
    "name": ("name", "name"),
```

```
        "city": ("city", "title")
}
df_a.rows.append(df_b, names_map=names_map).print()
```

This gives us the following result:

```
id_number    name         city
(int64)      (object)     (object)

-----------  -----------  ----------
143          Alice        Plymouth
225          Bob          Bradford
545          Charlie      Norwich
765          Dan          Bath
329          Erin         Manchester
152          Frank        Ripon
```

In the preceding example, we're declaring a custom mapping before concatenating, setting all the values from the id column of df_a and the title columns of df_b to columns with different names in the resulting dataframe.

But what if we want to concatenate on the column axis? We'll look at this next.

Concatenating columns

By concatenating columns, we can get the columns of a dataframe and put them in another. This will simply put the columns from one dataset into another, keeping the same indices for the values of both datasets.

To concatenate the columns from one dataset into another, you can simply call df.cols.concat(df_other), as follows:

```
df_a = op.create.dataframe({
    "id": [143, 225, 545],
    "name": ["Alice", "Bob", "Charlie"],
    "city": ["Plymouth", "Bradford", "Norwich"]
})
df_b = op.create.dataframe({
    "age": [25, 35, 45],
```

```
        "placeholder": ["foo", "bar", "baz"]
})
df_a.cols.concat(df_b).print()
```

This results in the following output:

id	name	city	age	placeholder
(int64)	(object)	(object)	(int64)	(object)
---------	---------	---------	---------	-------------
143	Alice	Plymouth	25	foo
225	Bob	Bradford	35	bar
545	Charlie	Norwich	45	baz

You can filter the columns of each dataset using `cols.select`, as follows:

```
df_a.cols.select(["id", "name"]).cols.concat(df_b.cols.select([
"age"])) .print()
```

This results in the following output:

id	name	age
(int64)	(object)	(int64)
---------	---------	---------
143	Alice	25
225	Bob	35
545	Charlie	45

When our data is not organized correctly, it can create rows that pairs values incorrectly. In that case, we can use a common column in both datasets that tells us what value corresponds to what row. In this case, we can use an operation called join.

Joining data

The join operation is used to merge entries from a data source to another using a common column as a key to pair the data correctly. The concept of joining is commonly seen in database technologies, in which we also see the different types of joins, such as inner join, outer join, left join, and right join. These joins are better represented in the following diagram:

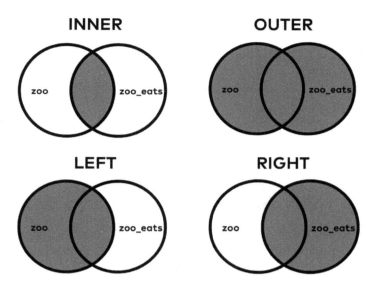

Figure 4.1 – Inner, outer, left, and right joins

When joining data, we must identify the key column for both dataframes. Let's look at an example:

```
df_a = op.create.dataframe({
    "id": [143, 225, 545, 765, 152],
    "name": ["Alice", "Bob", "Charlie", "Dan", "Frank"]
})
df_b = op.create.dataframe({
    "id": [225, 545, 765, 152, 329],
    "city": ["Bradford", "Norwich", "Bath", "Ripon",
"Manchester"],
    "placeholder": ["BRA", "NOR", "BAT", "RIP", "MAN"]
})
```

In both datasets, we have a column called `"id"` with equal values but in a different order. Calling `df.cols.join` requires us to input the on argument with the title of the column, as follows:

```
df_a.cols.join(df_b, how="outer", on="id").print()
```

We'll get the following result:

id (object)	name (object)	city (object)	placeholder (object)
143	Alice	nan	
225	Bob	Bradford	BRA
545	Charlie	Norwich	NOR
765	Dan	Bath	BAT
152	Frank	Ripon	RIP
329	nan	Manchester	MAN

As we can see, even when neither dataset is sorted, the result contains the correct data. As for **concat**, all the missing information is filled with nan. As we can see, we provided the "outer" to how, which means it will include all the values from both datasets.

By default, the join function uses "left" in the how argument. Let's look at an example of a left join (not including a value to that argument) so that we can omit the nan value corresponding to "name" in the last row:

```
df_a.cols.join(df_b, on="id").print()
```

We'll get the following result:

id (object)	name (object)	city (object)	placeholder (object)
143	Alice	nan	
225	Bob	Bradford	BRA
545	Charlie	Norwich	NOR
765	Dan	Bath	BAT
152	Frank	Ripon	RIP

However, there are still nan values on the city and placeholder columns. To omit them, we can use inner on how, which is another kind of join operation that will omit all the missing rows in both datasets:

```
df_a.cols.join(df_b, on="id", how="inner").print()
```

We'll get the following result:

```
id            name          city          placeholder
(object)      (object)      (object)      (object)

----------    ----------    -----------   ---------------

225           Bob           Bradford      BRA

545           Charlie       Norwich       NOR

765           Dan           Bath          BAT

152           Frank         Ripon         RIP
```

When the key columns of each dataframe have different names, you can use `left_on` and `right_on` instead of the `on` argument, like so:

```
df_a.cols.join(df_b, left_on="id_a", right_on="id_b")
```

The result will be ordered by default; that is, putting the key columns first, then the columns from the left, and then the columns from the right, as shown in the first example. If want to put the key column between the columns from the left dataset and the columns from the right dataset, you'll want to pass `True` to `key_middle`:

```
df_a.cols.join(df_b, left_on="id_a", key_middle=True).cols.
names()
```

In the preceding example, we're joining the datasets and getting the names of the columns in the resulting dataset, which results in the following:

```
["name", "city", "id", "placeholder"]
```

By default, the order of the columns in the resulting dataset will be `["id", "name", "city", "placeholder"]` instead.

Now that we know how to merge two datasets in various ways, let's learn how to reshape and pivot a dataframe that requires such transformations.

Reshaping and pivoting

In some cases, you'll want to make some more radical transformations to your dataset. In this section, we'll learn how to reshape an Optimus DataFrame in various ways, including pivoting, staking, and melting.

Pivoting

Pivoting is the process of reshaping stacked data into a new dataframe with simpler and less detailed data. It involves using the data of a column of choice and using it as column labels. Then, you must use one or more columns to group the data and calculate its values with the preferred summarization or aggregation over the rest of the data. The following is an example of this:

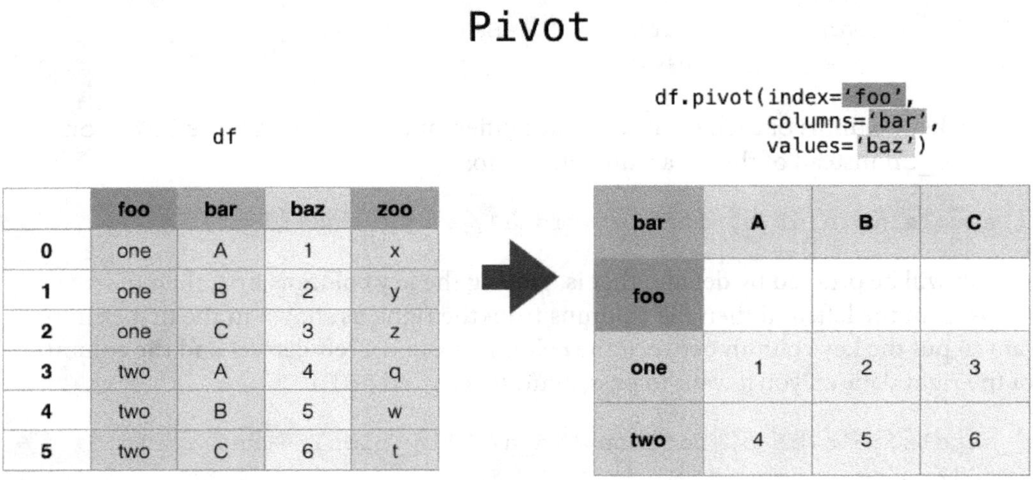

Figure 4.2 – How pivoting works

Let's look at an example of a dataset of sales that was made in a short period of time:

```
df = op.create.dataframe({
    "date": ["1/1/21", "1/1/21", "1/2/21", "1/2/21", "1/3/21",
"1/3/21", "1/3/21", "1/3/21", "1/3/21"],
    "product": ["Coffee", "Coffee", "Tea", "Coffee", "Tea",
"Coffee", "Tea", "Tea", "Coffee"],
    "size": ["big", "big", "big", "big", "big", "small",
"small", "small", "small"],
    "price": [1.5, 1.5, 2, 1.5, 2, 1, 1.25, 1.25, 1]
})
```

Our dataset looks like this:

date	product	size	price
(object)	(object)	(object)	(float64)
----------	----------	----------	----------

1/1/21	Coffee	big	1.5
1/1/21	Coffee	big	1.5
1/2/21	Tea	big	2
1/2/21	Coffee	big	1.5
1/3/21	Tea	big	2
1/3/21	Coffee	small	1
1/3/21	Tea	small	1.25
1/3/21	Tea	small	1.25
1/3/21	Coffee	small	1

Now, let's count how many sales per product were made each day:

```
df.pivot("date", groupby="product").print()
```

We'll get the following output:

| product | 1/1/21 | 1/2/21 | 1/3/21 |
(object)	(int64)	(int64)	(int64)
Coffee	2	1	2
Tea	0	1	3

As we can see, the default aggregation that was made is counting. We can explicitly call it using the agg argument, which requires a tuple with the name of the aggregation and the column to be calculated, as follows:

```
df.pivot("date", groupby="product", agg=("count", "date")).
print()
```

The preceding code will get us the same output we received previously.

Now, let's make use of the agg argument for other purposes:

```
df.pivot("date", groupby="product", agg=("common", "size")).
print()
```

We'll get the following output:

| product | 1/1/21 | 1/2/21 | 1/3/21 |
(object)	(object)	(object)	(object)

Coffee	big	small	
Tea	nan	big	['big', 'small']

In the previous example, we are obtaining the most common size of each product from the sales that were made on each date.

In the following example, we're looking at the mean of the prices of each product instead:

```
df.pivot("date", groupby="product", agg=("mean", "price")).
print()
```

We'll get the following output:

product (object)	1/1/21 (float64)	1/2/21 (float64)	1/3/21 (float64)
Coffee	1.5	2	1
Tea	nan	1.5	1.5

It's also possible to group by various columns instead of one, as follows:

```
df.pivot("date", groupby=["product", "size"]).print()
```

We'll get the following output:

product (object)	size (object)	1/1/21 (int64)	1/2/21 (int64)	1/3/21 (int64)
Coffee	big	2	1	0
Coffee	small	0	0	2
Tea	big	0	1	1
Tea	small	0	0	2

Pivoting is a special case of the inverse process of stacking. Let's learn what stacking is and how we can use it.

Stacking

When stacking a dataset, the column labels will be passed to the values in a dummy column and moved to match the column values, as shown here:

Stack

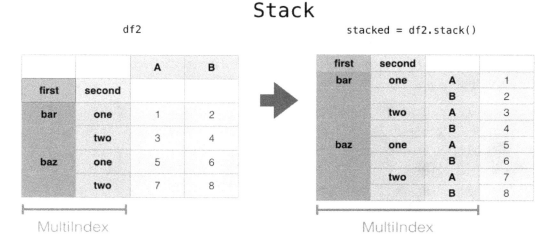

Figure 4.3 – How stacking works

To use stacking, we must know which column or columns represent the index in the dataframe. The result of using this function will give us a dataframe with a new index that contains the names of the rest of the columns.

Let's look at an example of an inventory, similar to the previous example:

```
df = op.create.dataframe({
    "product": ["Coffee", "Coffee", "Tea", "Tea"],
    "size": ["big", "small", "big", "small"],
    "price": [1.5, 1, 2, 1.25],
    "cost": [0.24, 0.2, 0.32, 0.3]
})
```

This is what our dataset looks like:

product (object)	size (object)	price (float64)	cost (float64)
Coffee	big	1.5	0.24
Coffee	small	1	0.2
Tea	big	2	0.32
Tea	small	1.25	0.3

To stack this, we need to call `df.stack`, as follows:

```
df.stack(index=["product", "size"]).print()
```

This will identify the first two columns as indices of the dataset. We'll see the following output:

product (object)	size (object)	variable (object)	value (float64)
Coffee	big	price	1.5
Coffee	big	cost	0.24
Coffee	small	price	1
Coffee	small	cost	0.2
Tea	big	price	2
Tea	big	cost	0.32
Tea	small	price	1.25
Tea	small	cost	0.3

The resulting dataset will require a name for each new column. By default, it will assign `"variable"` and `"value"`, but these names can be passed to the function, as follows:

```
df.stack(index=["product", "size"], col_name="foo", value_
name="bar").print()
```

This will identify the first two columns as indices of the dataset. We'll see the following output:

product (object)	size (object)	foo (object)	bar (float64)
Coffee	big	price	1.5
Coffee	big	cost	0.24
Coffee	small	price	1
Coffee	small	cost	0.2
Tea	big	price	2
Tea	big	cost	0.32
Tea	small	price	1.25
Tea	small	cost	0.3

So far, we've learned about pivoting and stacking and clarified that pivoting is a special case of the inverse process of stacking, which is called unstacking. We'll learn about this next.

Unstacking

Unstacking is the opposite of stacking. So, in this scenario, we must find out which column or columns represents the index in the dataframe. Based on that and the level that's passed to `df.unstack`, that index column will be transformed into columns:

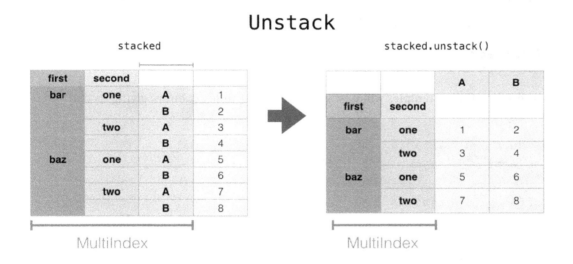

Figure 4.4 – How unstacking works

Let's see the result of unstacking the previously stacked dataframe:

```
df.unstack(index=["product", "size", "variable"]).print()
```

We'll get the following output:

product	size	cost	price
(object)	(object)	(float64)	(float64)
Coffee	big	0.24	1.5
Coffee	small	0.2	1
Tea	big	0.32	2
Tea	small	0.3	1.25

To pass which level you want to unstack, use the `level` argument, as follows:

```
df.unstack(index=["product", "size", "variable"], level=1).
print()
```

Negative numbers will also work:

```
df.unstack(index=["product", "size", "variable"], level=-2).
print()
```

In the preceding code, we are unstacking the second to last column in the multi-index.

Also, the name of the column can be passed to `level` too:

```
df.unstack(index=["product", "size", "variable"],
level="size").print()
```

The preceding three examples will unstack the `"size"` index column, giving us the following result:

product	variable	big	small
(object)	(object)	(float64)	(float64)
---------	---------	---------	---------
Coffee	cost	0.24	0.2
Coffee	price	1.5	1
Tea	cost	0.32	0.3
Tea	price	2	1.25

In this section, we learned about unstacking and how it relates to pivoting and stacking. Now, let's dig into a similar transformation that doesn't rely on explicit indices but on columns we can set as identifiers.

Melting

Melting is used to change the dataframe format from wide to long, similar to stack, but without indices. On the resulting dataframe, one or more columns will work as identifiers. The rest will be the values of the previous column (named `"values"` by default) and the identifier of those values (named `"variable"`). You can exclude columns in this function. You can see how melting works in the following diagram:

Melt

df3   df3.melt(id_vars=['first', 'last'])

	first	last	height	weight
0	John	Doe	5.5	130
1	Mary	Bo	6.0	150

	first	last	variable	value
0	John	Doe	height	5.5
1	Mary	Bo	height	6.0
2	John	Doe	weight	130
3	Mary	Bo	weight	150

Figure 4.5 – How melting works

To use it, the minimal arguments that are required are just the identifier columns:

```
df.melt(id_cols=["product","size"]).print()
```

In the preceding code, we just passed the identifier columns, but you can also explicitly pass the values you want to get, as follows:

```
df.melt(id_cols=["product","size"], value_cols=["price",
"cost"]).print()
```

The preceding code will return the following output:

product (object)	size (object)	variable (object)	value (float64)
Coffee	big	price	1.5
Coffee	small	price	1
Tea	big	price	2
Tea	small	price	1.25
Coffee	big	cost	0.24
Coffee	small	cost	0.2
Tea	big	cost	0.32
Tea	small	cost	0.3

To define the names of the resulting columns, you can use `var_name` and `value_name`:

```
df.melt(id_cols=["product", "size"], var_name="foo", value_
name="bar").print()
```

This will result in the following output:

product	size	foo	bar
(object)	(object)	(object)	(float64)
----------	----------	----------	-----------
Coffee	big	price	1.5
Coffee	small	price	1
Tea	big	price	2
Tea	small	price	1.25
Coffee	big	cost	0.24
Coffee	small	cost	0.2
Tea	big	cost	0.32
Tea	small	cost	0.3

As we can see, we can reduce the size of a dataset by using some of the transformations we just learned how to apply, but if you're coming from the relational database world, then you already know about aggregations, which can reduce drastically the size of a dataset but cause them to lose resolution.

Aggregations

Aggregations are functions where the values of multiple rows are grouped to form a single value.

Optimus comes with two ways to apply aggregations.

- Using the `cols` accessor
- Using the `agg` function

Let's look at both.

The cols accessor

For example, let's say you want to calculate the minimum value of a column. In Optimus, you can do something like this:

```
from optimus import Optimus
op = Optimus("pandas")
df = op.load.file("foo.csv")
df.print()
```

This will print the following output:

```
name         job              id
(object)     (object)       (int64)
----------   ----------    ---------
optimus      Leader           1
optimus      Espionage        2
bumblebee    1                3
bumblebee    3                4
```

To get the minimum value, we can use the following code:

```
df.cols.min("id")
```

You will get the number 1 as output:

```
1
```

You can also pass a list of columns, like so:

```
df.cols.min(["id","name"])
```

You will receive the following output:

```
{'id': 1, 'name': 'bumblebee'}
```

Here, three things happened:

- In contrast, when passing one column here, you don't just get the value – you get a Python dictionary containing the column's name and its value.
- Unlike traditional dataframe technologies that return a dataframe object, Optimus returns a Python dictionary.

This is useful for getting a single aggregation from our dataframe, but if you want to get a set of aggregations, you can use a different method, as we'll see next.

The agg method

Now, let's explore the agg dataframe method. With this method, you can calculate one or multiple aggregations at the same time.

For example, if you want to calculate the minimum aggregation, you can call the min method while passing {"id": "min"}, as shown here:

```
df.agg({"id": "min"})
```

This will return the following output:

```
1
```

Now, suppose that you want to apply multiple aggregations, such as min and max:

```
df.agg({"name": "min", "id": "max"})
```

This will return the min value of the name column and the max value of the id column:

```
{'name': 'bumblebee', 'id': 4}
```

Now that you know how to apply aggregation, here is the full list of aggregations supported by Optimus. It is very important to consider that some engines can make some aggregations in parallel. So, for example, you can't calculate two aggregations using cols, like so:

```
print(df.cols.std("id"))
print(df.cols.min("id"))
```

However, you can use the `agg` function to pass the two parameters at the same time, like so:

```
df.agg({"id":"std","name":min})
```

Dataframe engines are evolving every day. If you want to dig into which functions can be parallelized, go to the relevant documentation.

To finish this section, here is a full list of all the aggregations supported by Optimus:

Function	Description
Count	Number of non-null observations
Sum	Sum of values
Mean	Mean of values
Mad	Mean absolute deviation
Median	Arithmetic median of values
Min	Minimum
Max	Maximum
Mode	Mode
Abs	Absolute value
Prod	Product of values
Std	Unbiased standard deviation
Var	Unbiased variance
Skew	Unbiased skewness (third moment)
Kurt	Unbiased kurtosis (fourth moment)
Quantile	Sample quantile (value as a %)

Table 4.1 – Aggregation supported by Optimus

These aggregations will be performed on the whole dataset, without any groups being taken into account. Now, let's learn what to do if we want to limit the results to some groups.

Aggregating and grouping

Is a common use case to want to calculate the minimum, maximum, or any other aggregation in a dataset while considering a common set of values in another column. Here, we can use a practice called grouping. Let's try to explain this concept using the following diagram:

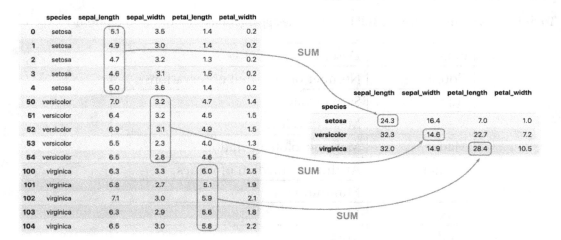

Figure 4.6 – Sum values applying grouping

In the previous example, we are summing the sepal_lenght values that are from the same species. Now, let's learn how to use grouping and aggregation in Optimus.

For example, we have the following dataframe:

```
df = op.load.file("foo.csv")
df.print()
```

The dataframe contains the following data:

name	job	id
(object)	(object)	(int64)
----------	----------	---------
optimus	Leader	1
optimus	Espionage	2
bumblebee	1	3
bumblebee	3	4

Perhaps you want to calculate the minimum `id` of the `optimus` and `bumblebee` values in the column `name` column. For this, the `agg()` function supports the `groupby` param, which will use the common values in the `name` column to calculate the minimum value in the `id` column. Let's see how it works:

```
A = df.agg({"id":min}, groupby="name")
```

This will return a Python dictionary containing the keys for the group name and the values for the minimum value of the group:

```
{'bumblebee': 3, 'optimus': 1}
```

After getting the Python dictionary, you can access the Bumblebee values just by using the following command:

```
A["bumblebee"]
```

This will return an integer representing the min value for the `"bumblebee"` group:

```
3
```

As you can see, the result is in a Python dictionary. The size of this result may vary, depending on how big your dataset is, so be aware of that before getting any results. You may want to request the output of this call in a dataframe instead, which you can do by passing `dataframe` into the output argument.

Summary

In this chapter, we learned about a lot of functions we can use to concatenate, merge, join, aggregate, and group data. All these functions will give you a lot of power to wrangle your data and shape it the way you need in order to get insights from it, reformat your dataset for different uses, and combine multiple datasets to get a better resolution of your data in the same dataset.

In the next chapter, we will learn about the functions that Optimus provides for profiling data so that we can get a better picture of it.

5
Data Visualization and Profiling

When you are transforming data, you usually need to explore your data in order to get a good understanding of how you can shape it to get insights from it. You may need to check for missing values, ensure consistency within a column, obtain a count of unique values, plot a histogram, get the top *n* values, or produce descriptive analytics. Optimus gives us tools to make all this and more happen.

In this chapter, we will deep dive into the profilers and their data types that we saw in *Chapter 3, Data Wrangling,* and see how we can fully take advantage of this feature to perform operations with specific data to set, drop, or replace values as you require.

Optimus can also give information about the quality of the data and provides the tools to process and transform our data easily.

The topics we will be covering in this chapter are as follows:

- Data quality
- Exploratory data analysis
- Data profiling
- Cache and flushing

Technical requirements

Optimus can work with multiple backend technologies to process data, including GPUs. With GPUs, Optimus uses **RAPIDS**, which needs an NVIDIA card. For more info about the requirements, please go to the *GPU configuration* section in *Chapter 1, Hi Optimus!*.

You can find all the code in this chapter at `https://github.com/ PacktPublishing/Data-Processing-with-Optimus`.

Data quality

In Optimus, we call the process of counting the number of values in a column that match a specific profiler data type **data quality**. For example, if the profiler data type in a column is URL, Optimus will count the number of values in a column that do the following:

- Match the URL format, such as `"google.com"`.
- Do NOT match the URL format, such as `"google"`.
- It will also count the null values.

Optimus has many data types in the profiler, which are inferred with a combination of regular expressions and number type detection. For reference, in the following table, we list the profiler data types and the Python data types:

Profiler Data type	Python Data type
Integer	String or integer
Floats	String or float
Strings	String
Email	String
URL	String
Gender	String
Boolean	String or Boolean
Zip Code	String or integer
Credit card number	String or integer
Date and time	String or datetime
Object	Python dictionary
Array	Python list
Phone number	String or integer
Social security number	String or integer
HTTP code	String or integer

Figure 5.1 – Optimus profiler datatypes

These data types are inferred when you run the profiler. Also, you can change the profiler if you are sure that a profiler datatype should have a specific data type:

```
from optimus import Optimus
op = Optimus("pandas")
df = op.load.file("foo.csv")
df.cols.quality()
```

`df.cols.quality` will return the following:

```
{'name': {'match': 4,
  'missing': 0,
  'mismatch': 0,
  'profiler_dtype': {'dtype': 'str', 'categorical': True}},
'job': {'match': 2,
  'missing': 0,
  'mismatch': 2,
  'profiler_dtype': {'dtype': 'int', 'categorical': True}},
'id': {'match': 4,
  'missing': 0,
  'mismatch': 0,
  'profiler_dtype': {'dtype': 'int', 'categorical': True}}}
```

The `quality` method returns a dictionary, with the column names as the keys, containing matches, mismatches, missing values, the name of the source file (if applicable), and `profiler_dtype`, which is the abstract type inferred by Optimus.

The first time you run the profiler, the data types of the columns are inferred using a sample of the dataset. Let's see a case in which we want to change the inferred data type for a column:

```
df = df.cols.set_dtype("salary", "int")
```

In the preceding code, we're changing the inferred data type of the `salary` column. If we get the profiler or get the data quality stats for that column, we'll get different results afterward. If we want to change the data type of multiple columns, we can call `df.cols.dtype`, passing a dictionary as follows:

```
df = df.cols.set_dtype({"salary": "int", "age": "str"})
```

In this case, we're setting a column to be taken as a string. This may result in fewer mismatches in the data quality.

There are special data types that are internally treated as a string but are constrained by a format, such as emails, URLs, and some datetime values.

In other cases, internally the values of datetime columns could be native datetime types (if the selected engine supports it).

It is also possible to expand the data types supported by Optimus. We will see more about this topic in the following chapters.

By setting a different data type, we're able to check the quality of our data more precisely. Let's see more about it.

Handling matches, mismatches, and nulls

In Optimus, to identify the values of each row that meet a given condition we use **masks**, which are simply rows of Boolean values that tell us if each value meets that condition. For example, when obtaining the null mask of a column with some null values, this mask will have the value **False** in all the rows except those with null values, which will have **True**:

```
df = op.create.dataframe({"numbers": [1, 2, None, 4]})
df.mask.missing("numbers").print()
```

We will obtain the following result:

```
  numbers
  (bool)
 ---------
       0
       0
       1
       0
```

This mask is useful to perform different row operations, such as row filtering or value replacing. We will see more about that later; first, we will see how to handle matches and mismatches using masks.

Let's say we have the following dataset:

```
df = op.create.dataframe({"numbers": [1, 2, "Hello", 4,
"World"]})
numbers
(object)
----------
1
2
Hello
4
World
```

If we want to know which values match the predominant data type (in this case, int) we use mask.match, passing "int" as the second argument:

```
df.mask.match("numbers", "int").print()
```

We will get the following output:

```
    numbers
    (bool)
---------
         1
         1
         0
         1
         0
```

To filter using this mask, we can use select or drop, passing the mask as the first argument:

```
df.rows.select(df.mask.match("numbers", "int")).print()
```

We will get the following output:

```
    numbers
    (object)
----------
         1
```

| 2 |
| 4 |

And we'll drop the row now:

```
df.rows.drop(df.mask.match("numbers", "int")).print()
```

We will get the following output:

```
numbers
(object)
----------
Hello
World
```

To replace the values using mask, we can do the following:

```
df.cols.set("numbers", value=0, where=df.mask.
mismatch("numbers", "int")).print()
```

In the preceding example, we're using mismatch instead of match. This enables us to replace all the values that aren't numbers in a column. The result of this is as follows:

```
  numbers
  (object)
  ----------
        1
        2
        0
        4
        0
```

By not passing a type into the mismatch method on mask, Optimus will use the data type available inferred previously, if it's available:

```
df.cols.dtypes("numbers")
df.mask.mismatch("numbers")
```

The preceding code will behave the same as the following:

```
df.cols.dtypes("numbers")
df.mask.mismatch("numbers", df["numbers"].profile.dtypes() )
```

That means we can simply call the following code and still get the same result:

```
df.cols.set("numbers", value=0, where=df.mask.
mismatch("numbers"))
```

We'll also get the following output:

```
    numbers
    (object)
    ----------
            1
            2
            0
            4
            0
```

This is helpful if we don't know what type should be used in a column and we have this information cached in our dataset.

We have learned how to handle data by its quality and how to clean our data. Once it's clean we can use this data to get some statistics. Let's learn how!

Exploratory data analysis

Exploratory Data Analysis (EDA) is a crucial step when you start exploring your data. It can give you an overall overview of its main characteristics, such as minimum and maximum values, as well as mean and median values. Also, it can help you to detect patterns, data inconsistencies, and outliers.

One of the first steps when exploring your data is to apply EDA techniques so you can get a better understanding of the data you want to process. The main goals of applying this technique are as follows:

- To maximize insight into a dataset
- To uncover the underlying structure
- To extract important variables
- To detect outliers and anomalies

There are four ways in which we can categorize EDA:

- **Single variable, non-graphical**: Here, the data analysis is applied to just one variable. The main purpose of univariate analysis is to describe the data and find patterns that exist within it.

- **Single variable, graphical**: Graphical methods for single variables can be a very intuitive way to explore your columns. Some of the plots available in Optimus are histograms, frequency charts, and box plots.

- **Multi-variable, non-graphical**: When you want to analyze the relationship between multiple variables you can rely on methods such as cross-tabulation or statistics.

- **Multi-variable, graphical**: These methods let you explore relationships between multiple variables in a graphical way. Optimus can help here with scatter plots and heat maps.

As you can see, in Optimus you can easily calculate almost every statistic you will need to know your data in depth.

Before diving into some examples, let's load a dataframe that resembles a store inventory:

```
from optimus import Optimus
op = Optimus("pandas")
df = op.load.file("store.csv")
df.print(10, ["name", "code"])
```

In the preceding code, we're calling `print` but, in this case, we are requesting the first ten rows of this dataset, and just two columns. This will print the following:

id (int64)	name (object)	code (object)	price (float64)
1	pants	L15	173.47
2	shoes	SH	69.99
3	shirt	RG30	30
4	pants	J10	34.99
5	pants	JG15	132.99
6	shoes	B	57.99
7	pants	JG20	179.99
8	pants	L20	95

```
 9   shirt        FT50              50
10   pants        JG15          169.99
```

But let's instead get an insight into whole columns by applying some of the available methods.

Single variable non-graphical methods

In Optimus, you can use certain methods to get non-graphical insights into the data. Let's see some of them.

To calculate the minimum value in a column, use the following:

```
df.cols.min("id")
```

This will return the following output:

```
1
```

If you also want to calculate the maximum value in a column you can use max, as shown here:

```
df.cols.max("id")
```

This will return the following:

```
504
```

On the other hand, if you want to calculate the mode, which is the most common value in the columns, you can use the following:

```
df.cols.mode("price")
```

This will return a dictionary or a single value depending on the data:

```
50.0
```

To calculate the median value in a column, use this:

```
df.cols.median("price")
```

This will return a numeric value:

```
104.99
```

To calculate the interquartile range, which is the range between Q1 and Q3, use this:

```
df.cols.iqr("id")
```

We'll get the following value:

```
130.01
```

To calculate the mean in the column, use this:

```
df.cols.mean("id")
```

This will return the following value:

```
121.30525793650794
```

Also, to calculate the standard deviation in the column we can use std:

```
df.cols.std("price")
```

We'll get a numeric value:

```
93.16652086384731
```

To calculate the variance in the column, use this:

```
df.cols.var("price")
```

We'll get the following value:

```
8680.000609873696
```

You can also calculate the skewness. This will tell you if the probability distribution is skewed to left or right:

```
df.data["price"].skew()
```

We'll get a numeric value:

```
1.0015117495305208
```

And for the kurtosis, which is a measure of the "tailedness" of the probability distribution, use this:

```
df.cols.kurtosis("price")
```

This will return the following value:

```
0.45556375186033016
```

We can also count some values by their possible properties. Let's see some of them.

To count all the zeros in a column you can use the following:

```
df.cols.count_zeros("discount")
```

We'll get the number of zeroes in `"discount"`:

```
294
```

To count all the null values in a column you can use the following:

```
df.cols.count_nulls("discount")
```

We'll get the following integer:

```
0
```

To count all the blank values in a column you can use the following:

```
df.cols.count_na("discount")
```

We'll get an integer value:

```
0
```

To count all the unique values in a column you can use the following:

```
df.cols.count_uniques("price")
```

This will return an integer with all the unique values in `"price"`:

```
192
```

As we can see, you can easily get specific insights from a column by using any of these methods. Now let's see how you can graphically explore your data.

Single variable graphical methods

Graphical data inspection can be a very intuitive way to get an insight into your data.

Optimus uses `matplotlib` and `seaborn`, a couple of very useful plotting libraries. Also, remember that you can output your data in Python dictionary format and use the library that best suits your needs.

Now, let's chart some data from our previously loaded dataset.

Histogram

A histogram tells us how many values are in each number of slices of numeric data, for example, how many people are in certain age groups.

To get the histogram of a numeric column you can use the following:

```
df.cols.hist("id",5)
```

This will print a Python dictionary showing the lower and upper bounds and the value count between them:

```
{'hist': {'price': [
{'lower': 5.0, 'upper': 103.3675, 'count': 250},
{'lower': 103.3675, 'upper': 201.735, 'count': 179},
{'lower': 201.735, 'upper': 300.1025, 'count': 39},
{'lower': 300.1025, 'upper': 398.47, 'count': 36}
]}}
```

To plot a histogram, you can use the following:

```
df.plot.hist()
```

This will display the following output:

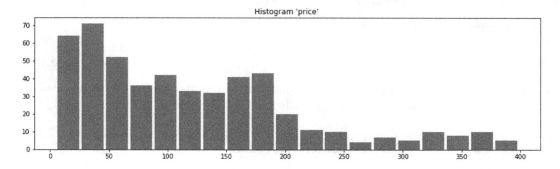

Figure 5.2 – Histogram chart generated using Optimus

This chart gives us an insight into the distribution of a numeric column in a numeric range. If the column is not numeric but categorical, you can create a frequency chart.

Frequency

With the `frequency` method, you can count how many times a value is present in one or multiple columns. By default, this is presented in descending order.

In Optimus to get the top five frequent values of any (or every) column, you can use the following:

```
df.cols.frequency("code", 5)
```

This will print a Python dictionary with the value and the count ordered in descending order:

```
{'frequency': {'code': {'values': [
{'value': 'JG15', 'count': 60},
{'value': 'JG10', 'count': 43},
{'value': 'SK', 'count': 37},
{'value': 'L15', 'count': 33},
{'value': 'J15', 'count': 32}
]}}}
```

You can plot a frequency chart using the following:

```
df.plot.frequency("code", 40)
```

This will display 40 bars like this:

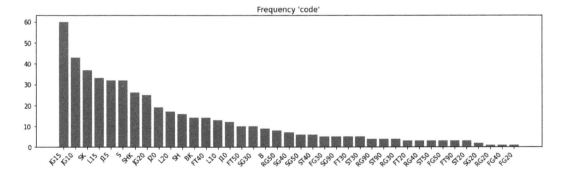

Figure 5.3 – Frequency chart generated using Optimus

Using this, you can find out the most frequent values in a column, as detailed as the number entered in the second argument.

Now, let's learn about a more advanced visualization for numeric data.

Box plot

A boxplot is a standardized way of displaying the distribution of data based on a five-number summary (minimum, first quartile (Q1), median, third quartile (Q3), and maximum).

Box plots are useful for analyzing numeric columns. Let's see how to get the data required to generate them:

```
df.cols.boxplot("price")
```

This will print a Python dictionary with all the data needed to print a boxplot, such as mean, median, first and third quartile, whiskers, and outlier points (also known as fliers):

```
{'price': {
    'mean': 121.30525793650794,
    'median': 104.99,
    'q1': 44.99,
    'q3': 175.0,
    'whisker_low': -150.02499999999998,
    'whisker_high': 370.015,
    'fliers': [
        {'price': 374.99},
        {'price': 395.0},
        {'price': 390.0},
        {'price': 395.0},
        {'price': 398.47},
        {'price': 380.0},
        {'price': 375.0}
    ],
    'label': 'price'
}}
```

To get a box plot with Optimus, you can use the following:

```
df.plot.box("age")
```

This will display the following plot:

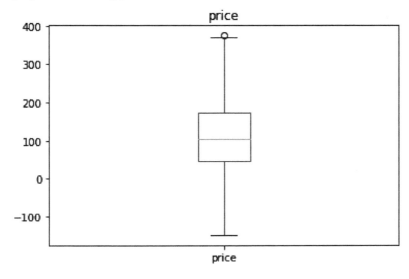

Figure 5.5 – Box plot generated using Optimus

This kind of plot can tell you about your outliers and what their values are. It can also tell you if your data is symmetrical, how tightly your data is grouped, and if and how your data is skewed.

Now that we have learned about plots that can help us explore columns individually, let's dig into other kinds of plots that may help us get a more general insight into multiple variables.

Multi-variable non-graphical methods

To learn about the whole dataset and how its variables relate to each other, you can use these types of methods. Let's discuss them.

Cross-tabulation

A cross-tabulation (also known as crosstab or contingency table) is a two-dimensional table that records the frequency of respondents that have specific characteristics described in the values of the table. It provides valuable information about the relationship between two variables.

To get a crosstab in Optimus, you can use the following:

```
df = op.create.dataframe(A=[18,21,62,44], B=[45,42,25,21])
df.cols.crosstab("A", "B")
```

This will output the following:

```
{21: {18: 0, 21: 0, 44: 1, 62: 0},
 25: {18: 0, 21: 0, 44: 0, 62: 1},
 42: {18: 0, 21: 1, 44: 0, 62: 0},
 45: {18: 1, 21: 0, 44: 0, 62: 0}}
```

You can also output this in a dataframe:

```
df.cols.crosstab("A", "B", output="dataframe")
```

This will print the following:

A (object)	21 (int64)	25 (int64)	42 (int64)	45 (int64)
18	0	0	0	1
21	0	0	1	0
44	1	0	0	0
62	0	1	0	0

As you can see, column A is maintained as an index. Let's see another way to see the relation between two columns.

Correlation

The correlation of two variables can convey how related two columns are. This value is represented in a numeric value between -1 and 1. A value of -1 means the values of each column are inversely correlated and a value of 1 means each column depends on the other, or they may even represent the same variable in different ways.

To get the correlation of two columns, you can use the following:

```
df = op.create.dataframe(A=[1,2,3,4], B=[4,5,0,7], C=[-1,-2,-5,-6])
df.cols.correlation(["A", "B"])
```

This will return a numeric value:

```
0.17541160386140586
```

If instead you pass more than two columns or even the whole dataset using "*", you'll get a dictionary representing the correlation matrix:

```
df.cols.correlation("*")
```

This will display the following:

```
{'A': {'A': 1.0, 'B': 0.17541160386140586, 'C':
-0.9761870601839528},
 'B': {'A': 0.17541160386140586, 'B': 1.0, 'C': 0.0},
 'C': {'A': -0.9761870601839528, 'B': 0.0, 'C': 1.0}}
```

Let's see how we can get graphical insights of the whole dataset.

Multi-variable graphical methods

A great way to get an insight into more than one column (or even the whole set of columns) is to use multi-variable graphical methods. Let's see some of them.

Heat map

A heat map plot is a type of plot in Cartesian space that displays information about two variables. It measures the magnitude of a phenomenon in two dimensions with a color variation.

To get a heat map in a Python dictionary format you can use the following:

```
df.cols.heatmap("fare")
```

This will return the following output:

```
{'frequency': {'name': {'values': [{'value': 'optimus',
'count': 2},
    {'value': 'bumblebee', 'count': 2}]},
  'job': {'values': [{'value': '1', 'count': 1},
    {'value': 'Leader', 'count': 1},
    {'value': 'Espionage', 'count': 1},
    {'value': '3', 'count': 1}]},
  'id': {'values': [{'value': '1', 'count': 1},
    {'value': '2', 'count': 1},
    {'value': '4', 'count': 1},
```

```
        {'value': '3', 'count': 1}]
}}}
```

To plot a heat map from a specific column, use this:

```
df.plot.heatmap("price", "id", 30, 30)
```

This will display the following plot:

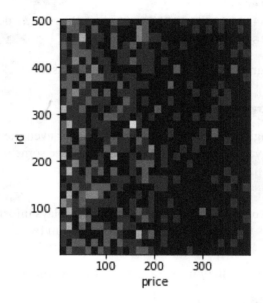

Figure 5.4 – Heat map generated using Optimus

As you can see, multiple values overlapping will be shown as a strong yellow color to represent how many points there are in the cluster.

Correlation matrix

A correlation matrix will show us the correlation coefficients between all the given columns. Let's load another dataset with more numeric columns before plotting:

```
df = op.load.file("titanic3.xls")
df.plot.correlation("*")
```

This will display a correlation matrix between every column in `df`:

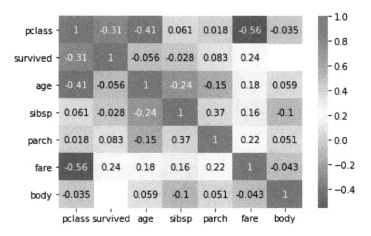

Figure 5.6 – Optimus correlation plot

Every pair of comparable columns has a color-coded value in it. This is useful for seeing patterns in our data.

Remember that Optimus can give you all this data in Python dictionary format, for example:

- Histograms using `df.cols.hist()`
- Frequency charts using `df.cols.frequency()`
- Box plots using `df.cols.boxplot()`
- Scatter plots using `df.cols.scatter()`

For a more general insight into the data, you can ask for a complete profile of the dataset. Let's check that out.

Data profiling

There is a handy function in Optimus called `profile` that returns useful stats about our dataset. Let's see how to use it:

```
df.profile(bins=5)
```

This code will return a dictionary:

```
{'columns': {'id': {'stats': {'match': 504,
    'missing': 0,
    'mismatch': 0,
    'profiler_dtype': {'dtype': 'int', 'categorical': True},
    'frequency': [{'value': 1, 'count': 1},
     {'value': 332, 'count': 1},
     {'value': 345, 'count': 1},
     {'value': 344, 'count': 1},
     {'value': 343, 'count': 1}],
    'count_uniques': 504},
   'dtype': 'int64'},
  'name': {'stats': {'match': 504,
    'missing': 0,
    'mismatch': 0,
    'profiler_dtype': {'dtype': 'str', 'categorical': True},
    'frequency': [{'value': 'pants', 'count': 254},
     {'value': 'shoes', 'count': 134},
     {'value': 'shirt', 'count': 116}],
    'count_uniques': 3},
   'dtype': 'object'},
  'code': {'stats': {'match': 504,
    'missing': 0,
    'mismatch': 0,
    'profiler_dtype': {'dtype': 'str', 'categorical': True},
    'frequency': [{'value': 'JG15', 'count': 60},
     {'value': 'JG10', 'count': 43},
     {'value': 'SK', 'count': 37},
     {'value': 'L15', 'count': 33},
     {'value': 'J15', 'count': 32}],
    'count_uniques': 39},
   'dtype': 'object'},
  'price': {'stats': {'match': 504,
    'missing': 0,
    'mismatch': 0,
```

```
        'profiler_dtype': {'dtype': 'decimal', 'categorical':
False},
        'hist': [{'lower': 5.0, 'upper': 103.3675, 'count': 250},
          {'lower': 103.3675, 'upper': 201.735, 'count': 179},
          {'lower': 201.735, 'upper': 300.1025, 'count': 39},
          {'lower': 300.1025, 'upper': 398.47, 'count': 36}]},
        'dtype': 'float64'},
      'discount': {'stats': {'match': 294,
        'missing': 0,
        'mismatch': 210,
        'profiler_dtype': {'dtype': 'int', 'categorical': True},
        'frequency': [{'value': '0', 'count': 294},
          {'value': '5%', 'count': 65},
          {'value': '20%', 'count': 63},
          {'value': '15%', 'count': 54},
          {'value': '50%', 'count': 16}],
        'count_uniques': 6},
      'dtype': 'object'}},
  'name': 'store.csv',
  'file_name': ['store.csv'],
  'summary': {'cols_count': 5,
    'rows_count': 504,
    'dtypes_list': ['float64', 'int64', 'object'],
    'total_count_dtypes': 3,
    'missing_count': 0,
    'p_missing': 0.0}
  }
```

With this Python dictionary, you can get info about specific columns and stats about the whole dataframe.

For dataframe stats, you can use `profile.summary()` to get the following:

- `cols_count`: Number columns in the dataframe
- `rows_count`: Number of rows in the dataframe
- `dtypes_list`: List of dtypes in the dataframe
- `total_count_dtypes`: Count of data types in the dataframe

- `missing_count`: Number of missing values in the dataframe
- `p_missing`: Percentage of missing values in the dataframe

Using `profile.columns()`, you can get info about every column in the dataframe. Inside this, you can access two keys, `stats` and `dtype`.

In `stats`, you can get this info:

- `match`: Number of values in the columns that match the `profiler_dtype`
- `missing`: Number of missing values
- `Mismatch`: Number of values in the column that do not match the `profiler_dtype`, excluding null values
- `profiler_dtype`: Datatype inferred by Optimus
- `Frequency`: Top *n* values in descending order
- `Hist`: Density of value in every bin
- `count_uniques`: Number of unique values

Optimus will calculate a frequency or a histogram depending on the datatype. It will calculate a histogram for numeric data types and a frequency for string datatypes.

All this displayed info can give us a quick insight into what's in a dataset, but by caching this metadata we can gain some time. Let's learn more about that.

Cache flushing

Exploring big data can be a very time-consuming process. You need to operate over a column, transform its data, check that the output is what you want, and compare it with data in another column, including its frequency, histogram, and descriptive analytics.

To help you accelerate your job, Optimus knows when it needs to recalculate the profiler stats so you do not have to wait, which can be very helpful if you are handling big data.

Internally, Optimus declares some Actions that trigger the column profile recalculation. To get the full list of Actions that will require a recalculation, you can use the following:

```
from optimus.helpers.constants import Actions
Actions.list()
```

`Actions.list()` will get us a Python list:

```
['profiler_dtype', 'lower', 'upper', 'proper', 'pad', 'trim',
'reverse', 'remove', 'left', 'right', 'mid', 'replace',
'fill_na', 'cast', 'is_na', 'z_score', 'nest', 'unnest',
'set', 'string_to_index', 'date_format', 'index_to_string',
'min_max_scaler', 'max_abs_scaler', 'apply_cols', 'impute',
'extract', 'abs', 'math', 'variance', 'slice', 'clip', 'drop',
'keep', 'cut', 'to_float', 'to_integer', 'to_boolean', 'to_
string', 'years', 'append', 'port', 'domain', 'domain_scheme',
'subdomain', 'host', 'domain_params', 'domain_path', 'email_
domain', 'email_user', 'select_row', 'drop_row', 'between_
drop', 'sort_row']
```

If you look at these names, some of them match the functions to process data that we have already used.

Summary

In this chapter, we learned how to get extract quality data from our data so we can apply a transformation to shape it and start getting quality stats, which can help us to understand the relations between the data and extract better insights.

Also, we saw how Optimus can plot this data to put it in a format that is easy to consume and understand.

Now that we know how to explore our data in depth, in the next chapter, we will learn how to apply string clustering techniques to easily find groups of different values that might be alternative representations of the same thing.

6
String Clustering

Frequently when wrangling data, you will find columns that look as though they have similar values, but they do not. To handle this task, Optimus gives you some handy techniques through which you can easily detect which strings are similar and group them, giving you some options that could point to the best value in the group. We will explore all these techniques in this chapter.

In this chapter, we will learn about the following topics:

- Exploring string clustering
- Key collision methods
- Phonetic encoding
- Nearest-neighbor methods
- Applying suggestions

Technical requirements

Optimus can work with multiple backend technologies to process data, including **graphics processing units (GPUs)**. For GPUs, Optimus uses the **Real-Time Automated Personnel Identification System (RAPIDS)**, which needs an NVIDIA card. For more information about the requirements, please go to the *GPU configuration* section in *Chapter 1, Hi Optimus!*.

You can find all the code for this chapter at `https://github.com/PacktPublishing/Data-Processing-with-Optimus`.

Exploring string clustering

String clustering may be one of the most underrated data-cleansing functions. To explain what string clustering is for, we can refer to the definition from *OpenRefine*, which says: *"find groups of different values that could be alternative representations of the same thing"*:

For example, you may have a column with values like this:

```
A
(object)
------------
Optimus
Optimus Prime
Prime
```

All these values can be represented as the same string since they reference the same thing—for example, `"Optimus"` or `"Optimus Prime"` are valid options, depending on the need. Optimus will give you the tools to apply different string-clustering methods, suggest a value that best represents what you want, and then replace the values to achieve a cohesive representation of the data.

Optimus gives us the possibility to use different string-clustering methods, from some fast and less accurate methods such as **fingerprinting** to more advanced ones such as **Levenshtein distance**. Which one you will use depends on your use case, the size of the data at hand, and the computing power available to you.

All these methods tend to be a combination of operations over strings already defined in Optimus, such as `lowercase` or `remove_special_symbols`. We will be explaining every method in detail so that you can have a better understanding of how they operate internally and know which one adapts better to your use case. We will also be explaining other terms related to the field—such as **n-gram** and **Levenshtein distance**—so that you can understand what is happening when using this feature.

In Optimus, we can divide the methods available for that practice into two groups: **key collision methods** and **nearest-neighbor methods**.

We will deep dive into the algorithms available in every group and how to use them to normalize similar values.

Key collision methods

Key collision methods are based on the idea of creating a reduced and meaningful representation of a value (a **key**) and putting equal ones together in buckets.

Optimus has implemented three methods that fall into this category: **fingerprinting**, **n-gram fingerprinting**, and **phonetic fingerprinting**.

Fingerprinting

A fingerprinting method is the least likely to generate false positives, which is why Optimus defaults to this.

Optimus implements the same algorithm as OpenRefine, an open source tool for working with messy data. The algorithm is described in the next code block.

The process that generates a key from a string value is outlined here and must be followed in this order:

1. Remove leading and trailing whitespace (for example, from " Optimus Prime" to "Optimus Prime").

2. Change all characters to their lowercase representation (for example, from "Optimus Prime" to "optimus prime").

3. Remove all punctuation and control characters (characters that help to give form to the text but cannot be seen, such as a tab or a carriage return, among others).

4. Update extended western characters with their **American Standard Code for Information Interchange** (**ASCII**) representation (for example, from "öptimus" to "optimus").

5. Divide the text string into individual tokens for every word after a whitespace (for example, from "optimus prime" to ["optimus", "prime"]).

6. Sort and remove duplicates in the tokens (for example, ["optimus", "prime"]. Because *o* is lower than *p*, there is no modification in the item order.

7. Finally, club the tokens together. The result would be ["optimus prime"].

Now that we know how fingerprinting methods work, let's look at an example using Optimus. First, we will create an Optimus dataframe, as follows:

```
df = op.create.dataframe({
    "A": ["optimus", "prime optimus", "prime", "bumblebee",
"megatron", "MEGATRON"],
```

```
    "B": [1,2,3,4,5,6]
})
```

Let's apply a fingerprinting method to get the clusters from `name`. To achieve that, we'll be using a method called `string_clustering`, passing the name of the algorithm in the second argument—in this case, `"fingerprint"`. The code is illustrated in the following snippet:

```
clusters = df.string_clustering("A", "fingerprint")
clusters
```

This would give us the following output:

```
{ 'A': { 'bumblebee': { 'cluster': 'bumblebee',
                        'suggestions': ['bumblebee'],
                        'suggestions_size': 1,
                        'total_count': 1},
         'megatron': { 'cluster': 'megatron',
                       'suggestions': ['megatron'],
                       'suggestions_size': 1,
                       'total_count': 2},
         'optimus': { 'cluster': 'optimus',
                      'suggestions': ['optimus'],
                      'suggestions_size': 1,
                      'total_count': 1},
         'optimus prime': { 'cluster': 'optimus prime',
                            'suggestions': ['optimus prime'],
                            'suggestions_size': 1,
                            'total_count': 1},
         'prime': { 'cluster': 'prime',
                    'suggestions': ['prime'],
                    'suggestions_size': 1,
                    'total_count': 1}
}}
```

In this result, we're getting the clusters in a dictionary. Because the structure of the result is the same for all the string-clustering methods later in this chapter, we'll explain the result we get in detail.

You can also apply a fingerprinting method over a specific column, like this:

```
print(df.cols.fingerprint("A"))
```

To transform and reduce column A to its fingerprint, we'll use the following code:

A (object)	B (int64)
optimus	1
optimus prime	2
prime	3
bumblebee	4
megatron	5
megatron	6

Here's a summary of this result:

- bumblebee, megatron, optimus, and prime have the same output.
- prime optimus is converted to optimus prime because the token is reordered.

N-gram fingerprinting

To understand this method, we should first talk about n-grams. An n-gram can be a sequence of *n* things. In this case, an n-gram is a sequence of characters—for example, a 2-gram of the optimus string is ["op", "pt", "ti", "im", "mu", "us"]. If you take a closer look, to build a 2-gram you need to take the two first characters of a string, then take the last character of the last two characters you took and add the next one, and repeat the process until you get to the end of the string.

> **Note**
> Optimus implements the same algorithm as OpenRefine, an open source tool for working with messy data. The algorithm is described in the next code block.

An n-gram fingerprinting algorithm works like this:

1. Changes all characters to their lowercase representation.
2. Removes all punctuation, whitespace, and control characters (characters that help to give form to the text but cannot be seen, such as a tab or a carriage return, among others).

3.   Creates all string n-grams.

4.   Sorts the n-grams and removes duplicates.

5.   Joins the sorted n-grams back together as a string.

6.   Converts extended western characters to their ASCII representation.

The whole idea is to separate a string into small chunks. This is useful in practice since there is no advantage of using big values for n-grams compared to a fingerprinting method. However, using 2-grams and 1-grams can find clusters that earlier methods couldn't, even with strings that have minor differences, although they do yield several false positives.

Let's look at an example here:

```
df = op.create.dataframe({
    "A": ["optimus", "optimus prime", "prime", "bumblebee",
"megatron", "MEGATRON"],
    "B": [1,2,3,4,5,6]
})
```

To get the string clusters of this dataframe, we can use the same `string_clustering` methods that we use to apply a fingerprinting algorithm and pass the name of the algorithm in the second argument, as follows:

```
clusters = df.string_clustering("name, "ngram_fingerprint")
clusters
```

This would give us the following output:

```
{ 'A': { 'bumblebee': { 'cluster': 'beblbuebeelembum',
                        'suggestions': ['bumblebee'],
                        'suggestions_size': 1,
                        'total_count': 1},
        'megatron': { 'cluster': 'ateggameonrotr',
                      'suggestions': ['megatron'],
                      'suggestions_size': 1,
                      'total_count': 2},
        'optimus': { 'cluster': 'immuoppttius',
                     'suggestions': ['optimus'],
                     'suggestions_size': 1,
                     'total_count': 1},
```

```
               'optimus prime': { 'cluster': 'immemuopprptrisptius',
                                  'suggestions': ['optimus prime'],
                                  'suggestions_size': 1,
                                  'total_count': 1},
               'prime': { 'cluster': 'immeprri',
                          'suggestions': ['prime'],
                          'suggestions_size': 1,
                          'total_count': 1}
}}
```

We get the preceding output because the regular fingerprinting algorithm doesn't support changes well in the order or the repetitions of the characters in its entries—for example, `"Krzysztof"`, `"Kryzysztof"`, and `"Krzystof"` have varying lengths and varying fingerprints but share the same 1-gram fingerprint because they use the same letters.

As with fingerprinting methods, you can also apply an n-gram fingerprint to a column. In the following case, we will apply a method to A and output the result to C:

```
print(df.cols.ngram_fingerprint("A", output_cols="C"))
```

This would print the following output:

A	C	B
(object)	(object)	(int64)
-------------	---------------------	---------
optimus	immuoppttius	1
optimus prime	immemuopprptrisptius	2
prime	immeprri	3
bumblebee	beblbuebeelembum	4
megatron	ateggameonrotr	5
megatron	ateggameonrotr	6

Phonetic encoding

Phonetic fingerprinting is a method to encode strings into a representation that better matches the way they are pronounced. The main goal is to bucket similar-sounding words and sentences. For example, `"Reuben Meza"` and `"Ruben Mesa"` share the same phonetic fingerprint for English pronunciation, but they have different fingerprints for both the preceding regular and n-gram fingerprinting methods, no matter the size of the n-gram.

There are several phonetic methods, such as the following:

- Soundex
- Metaphone
- Double Metaphone
- Match Rating Codex
- **New York State Identification and Intelligence System (NYSIIS)**

Let's review them in a couple of paragraphs, based on their performance.

Soundex

Soundex was created by Robert C. Russell and Margaret King Odell, was patented in 1918 and 1922, and is one of the most popular phonetic algorithms. It was originally designed for American English; however, it is available today in various language-specific versions such as French, German, and Hebrew, which are not present in Optimus.

The Soundex algorithm functions like this:

1. Hold on to the first letter of the name and remove all the other instances of a, e, i, o, u, y, h, and w.

2. Swap consonants with digits as follows (after the first letter):

Characters	Number
b, f, p, v	1
c, g, j, k, q, s, x, z	2
d, t	3
l	4
m, n	5
r	6

Figure 6.1 – Character-number replacement in Soundex

3. If two or more letters with the same number are beside each other, we will only keep the first letter. Any two letters that have the same number, but are separated by h or w, will be considered as a single number, and if they are separated by vowels, then they are considered twice. This rule goes for the first letter as well.

4. If there are less letters in your word, due to which only 2 numbers can be assigned, we will add zeros to the word until 3 numbers can be assigned. If you have more than 3 numbers, only the first three will be considered.

Now that we know how Soundex works, let's see it in action, as follows:

1. First, let's create a dataframe, like this:

```
df = op.create.dataframe({
    "A": ["optimus", "prime aptimus", "bumblebee",
"megatron", "MaGATRaN"],
    "B": [1,2,3,4,5,6]
})
```

2. To create a cluster using the soundex algorithm, we will use the following code:

```
df.string_clustering("A", "soundex")
```

3. To get a Python dictionary with suggestions, we will use the following code:

```
{ 'A': { 'bumblebee': { 'cluster': 'B514',
                        'suggestions': ['bumblebee'],
                        'suggestions_size': 1,
                        'total_count': 1},
         'megatron': { 'cluster': 'M236',
                       'suggestions': ['megatron',
'MEGATRON'],
                       'suggestions_size': 2,
                       'total_count': 2},
         'optimus': { 'cluster': 'O135',
                      'suggestions': ['optimus', 'optimus
prime'],
                      'suggestions_size': 2,
                      'total_count': 2},
         'prime': { 'cluster': 'P650',
                    'suggestions': ['prime'],
                    'suggestions_size': 1,
                    'total_count': 1}
}}
```

4. Now, let's see how to encode column A using Soundex. Here's the code to accomplish this:

```
print(df.cols.soundex("A", output_cols="C"))
```

This would print the following dataframe:

A	C	B
(object)	(object)	(int64)
-------------	----------	---------
optimus	O135	1
prime optimus	P651	2
prime aptimus	P651	3
bumblebee	B514	4
megatron	M236	5
MaGATRaN	M236	6

As we can see, similar-sounding strings were bucketed together. Looking at the results, we can note the following:

- `megatron` and `MaGATRaN` have the same phonetic fingerprint, `M236`, because it does not consider the consonants.

- `prime optimus` and `prime aptimus` have the same fingerprint, `P651`.

Metaphone

Metaphone is a phonetic algorithm created by Lawrence Phillips in 1990, and is used for assigning indexes to words using their English pronunciation. It improves on Soundex, by using various minor variances or inconsistent spelling and pronunciation errors, to generate a more accurate encoding, which then can be used with words or names that are similar to each other.

The algorithm itself is lengthy and consists of a bunch of comparisons and replacements. If you want to take a look, you can learn more about it at `https://en.wikipedia.org/wiki/Metaphone`.

In Optimus, you would apply string clustering using Metaphone like this:

```
df.string_clustering("A", "metaphone")
{ 'A': { 'bumblebee': { 'cluster': 'BMBLB',
                        'suggestions': ['bumblebee'],
                        'suggestions_size': 1,
                        'total_count': 1},
         'megatron': { 'cluster': 'MKTRN',
                       'suggestions': ['megatron', 'MEGATRON'],
```

```
                        'suggestions_size': 2,
                        'total_count': 2},
          'optimus': { 'cluster': 'OPTMS',
                        'suggestions': ['optimus'],
                        'suggestions_size': 1,
                        'total_count': 1},
    'optimus prime': { 'cluster': 'OPTMS PRM',
                        'suggestions': ['optimus prime'],
                        'suggestions_size': 1,
                        'total_count': 1},
            'prime': { 'cluster': 'PRM',
                        'suggestions': ['prime'],
                        'suggestions_size': 1,
                        'total_count': 1}
}}
```

To get a closer look, let's apply metaphone to the A column, as follows:

```
print(df.cols.metaphone("A"))
```

This would give us the following output:

```
A                B
(object)         (int64)
----------       ---------
OPTMS            1
PRM OPTMS        2
PRM APTMS        3
BMBLB            4
MKTRN            5
MKTRN            6
```

Looking at the results, we can note the following:

- Megatron and MaGATRaN have the same phonetic fingerprint, MKTRN.

- Because Metaphone can get differences in pronunciation, prime optimus and prime aptimus have different fingerprints, PRM OPTMS and PRM APTMS.

Double Metaphone

The Double Metaphone is the second generation which improves further on the Metaphone algorithm.

It is called Double as it can return either one or two code values for any string, a primary and a secondary one. This can be useful for unclear cases and also for variations in surnames under common ancestry.

Now, let's see an example of how to use the Double Metaphone method and how it can produce one or a pair of codes.

First, let's create a dataframe, as follows:

```
df = op.create.dataframe({
    "A": ["optimus prime", "prime optimus", "prime",
"bumblebee", "megatron", "MEGATRON", "argenis leon"],
    "B": [1,2,3,4,5,6,7]
})
```

To get the string clusters of this dataframe, we can use the same function as before, `string_clustering`, and pass the name of the algorithm in the second argument. The code to accomplish this is illustrated in the following snippet:

```
clusters = df.string_clustering("A, "double_metaphone")
clusters
```

This would give us the following output:

```
{ 'A': { 'argenis leon': { 'cluster': ('ARJNSLN', 'ARKNSLN'),
                           'suggestions': ['argenis leon'],
                           'suggestions_size': 1,
                           'total_count': 1},
        'bumblebee': { 'cluster': ('PMPLP', ''),
                      'suggestions': ['bumblebee'],
                      'suggestions_size': 1,
                      'total_count': 1},
        'megatron': { 'cluster': ('MKTRN', ''),
                     'suggestions': ['megatron', 'MEGATRON'],
                     'suggestions_size': 2,
                     'total_count': 2},
        'optimus prime': { 'cluster': ('APTMSPRM', ''),
```

```
                        'suggestions': ['optimus prime'],
                        'suggestions_size': 1,
                        'total_count': 1},
        'prime': { 'cluster': ('PRM', ''),
                   'suggestions': ['prime'],
                   'suggestions_size': 1,
                   'total_count': 1},
        'prime optimus': { 'cluster': ('PRMPTMS', ''),
                           'suggestions': ['prime optimus'],
                           'suggestions_size': 1,
                           'total_count': 1}
  }}
```

Because Double Metaphone produces a couple of values, to make the string clustering we calculate the double metaphone for every string and compare every string with all others to calculate which one is closer.

To calculate how close two strings are, Optimus takes the first value of every tuple and compares if they are equal. If not, the strings are not taken as similar and are not suggested.

Now, let's apply `double_metaphone` to column A to see what happens, as follows:

```
print(df.cols.double_metaphone("A"))
```

This would print the following output:

A	B
(object)	(int64)
('APTMSLN', '')	1
('PRMPTMS', '')	2
('PRM', '')	3
('PMPLP', '')	4
('MKTRN', '')	5
('MKTRN', '')	6
('ARJNSLN', 'ARKNSLN')	7

Key collision methods are very quick; however, they tend to be too strict or too lenient and have no way of tweaking how much difference between strings we want to handle.

We'll look into alternatives now.

Match Rating Codex

The **match rating approach (MRA)** was developed by Western Airlines in 1977. It can be used for encoding and comparing homophonous words that have the same pronunciation but different meaning, origin, or spelling.

The encoding rules just take the strings and transform them as follows:

- Remove all vowels unless the string begins with one.
- In the case of double consonants, remove the second one.
- Join the first three and last three letters to reduce the codex to six letters.

After encoding the string, we will apply comparison rules, as follows:

- If the length of both encoded strings differs by three characters or more, no similarity comparison is made.
- Obtain the minimum rating value by finding the length sum of the encoded strings, as follows:

Sum of Length	Minimum Rating
≤ 4	5
$4 < \text{sum} \leq 7$	4
$7 < \text{sum} \leq 11$	3
$= 12$	2

Figure 6.2 – String cluster method speed versus accuracy

- Remove any matching characters found from left to right.
- Find the similarity rating by subtracting the number of unmatched characters by 6 in the longer string.
- If the similarity rating is equal to or greater than the minimum rating, the match is considered good.
- A match is considered good if the similarity rating is equal to or more than the minimum rating.

To apply the Match Rating Codex phonetic method in Optimus, we will use the following code:

```
print(df.string_clustering("A", "match_rating_codex"))
```

This would print the following output:

```
{ 'A': { 'bumblebee': { 'cluster': 'BMBLB',
                        'suggestions': ['bumblebee'],
                        'suggestions_size': 1,
                        'total_count': 1},
         'megatron': { 'cluster': 'MGTRN',
                       'suggestions': ['megatron', 'MEGATRON'],
                       'suggestions_size': 2,
                       'total_count': 2},
         'optimus': { 'cluster': 'OPTMS',
                      'suggestions': ['optimus'],
                      'suggestions_size': 1,
                      'total_count': 1},
         'optimus prime': { 'cluster': 'OPTPRM',
                            'suggestions': ['optimus prime'],
                            'suggestions_size': 1,
                            'total_count': 1},
         'prime': { 'cluster': 'PRM',
                    'suggestions': ['prime'],
                    'suggestions_size': 1,
                    'total_count': 1}
}}
```

To get a better idea of how it works, let's apply `match_rating_codex` to column A and check the output, as follows:

```
print(df.cols.match_rating_codex("A"))
```

This would give us the following output:

A	C	B
(object)	(object)	(int64)
-------------	--------	-------
optimus	OPTMS	1
prime optimus	PRMTMS	2
prime aptimus	PRMTMS	3
bumblebee	BMBLB	4

| megatron | MGTRN | 5 |
| MaGATRaN | MGTRN | 6 |

In the first step, Match Rating Codex deletes the vowels, resulting in the following outcome:

- `prime optimus` and `prime aptimus` get the same value, `PRMTMS`.
- `MaGATRaN` and `megatron` get the same value, `MGTRN`.

NYSIIS

The NYSIIS phonetic method was developed in 1970 by the New York State Identification and Intelligence Center. It is similar to Soundex, meaning that when there are homophones, we have to match them by using indices for sound cues. The best part is the results; NYSIIS is more accurate than Soundex since it returns fewer surnames under the same code.

The algorithm itself is very lengthy. To get a detailed explanation, you can learn more about it at `https://en.wikipedia.org/wiki/New_York_State_Identification_and_Intelligence_System`.

To use string clustering using NYSIIS in Optimus, you would run the following code:

```
df.string_clustering("A", "nysiis")
```

This would give us the following clusters:

```
{ 'A': { 'bumblebee': { 'cluster': 'BANBLABY',
                        'suggestions': ['bumblebee'],
                        'suggestions_size': 1,
                        'total_count': 1},
         'megatron': { 'cluster': 'MAGATRAN',
                       'suggestions': ['megatron', 'MEGATRON'],
                       'suggestions_size': 2,
                       'total_count': 2},
         'optimus': { 'cluster': 'OPTAN',
```

```
                    'suggestions': ['optimus', 'optimus
prime'],
                    'suggestions_size': 2,
                    'total_count': 2},
        'prime': { 'cluster': 'PRAN',
                   'suggestions': ['prime'],
                   'suggestions_size': 1,
                   'total_count': 1}
}}
```

To apply match rating encoding to one or more columns, you can run the following code:

```
print(df.cols.nysiis("A"))
```

This would give us a result over column A, as illustrated here:

A	C	B
(object)	(object)	(int64)
optimus	OPTAN	1
prime optimus	PRAN	2
prime aptimus	PRAN	3
bumblebee	BANBLABY	4
megatron	MAGATRAN	5
MaGATRaN	MAGATRAN	6

From the result, we can see that these strings are homophones, so we get the following outcome:

- `prime optimus` and `prime aptimus` get PRAN.
- `Megatron` and `MaGATRaN` get MAGATRAN.

Now that we have explored key collision methods, let's see how nearest-neighbor methods work.

Nearest-neighbor methods

A nearest-neighbor method gives a parameter that represents the threshold of the distance between strings; any string pairs that have a distance value closer to the specified one will be grouped together, as illustrated in the following screenshot:

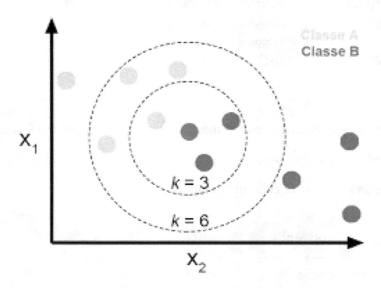

Figure 6.3 – Nearest neighbor with different radius values

Let's see some of those methods. For **k-Nearest Neighbor** (**kNN**) methods, Optimus implements Levenshtein distance. Let's see how this works.

Levenshtein distance

The Levenshtein distance between two words is calculated as the minimum number of single-character changes that need to be done to a word to convert it into another.

In this example, let's look at the necessary steps to transform a string, `"AABBCC"`, to `"ABZ"`.

Let's refer to `"AABBCC"` as String1 and `"ABZ"` as String2. We'll proceed as follows:

- First, delete `"A"` from String1 (`"AABBCC"` to `"ABBCC"`).
- Then, delete `"B"` from String1 (`"ABBCC"` to `"ABCC"`).
- Next, delete `"C"` from String1 (`"ABCC"` to `"ABC"`).
- Finally, substitute `"Z"` from String2 with `"C"` in String1 (`"ABC"` to `"ABZ"`).

Optimus will calculate the Levenshtein distance between all the strings in the columns and select the ones with the shortest distance, as follows:

```
df = op.create.dataframe({
    "name": ["John Doe", "alice", "alice", "John Doe", "álice",
    "john doe", "doe, john", "alice", "joohn dooe"]
})
```

To get the clusters of this dataframe, we can use the same function as before, string_clustering, and pass the name of the algorithm as the second argument, as follows:

```
clusters = df.string_clustering("name", "levenshtein")
clusters
```

This would give us the following output:

```
{
    "name": {
        "John Doe": {
            "suggestions": [
                "John Doe", "john doe", "doe, john", "joohn
dooe"
            ],
            "suggestions_size": 4,
            "total_count": 5
        },
        "alice": {
            "suggestions": ["alice", "álice"],
            "suggestions_size": 2,
            "total_count": 4
        }
    }
}
```

In this case, the method will suggest "John Doe", "john doe", "doe, john", and "joohn dooe" as being like "John Doe".

It's important to know that besides Levenshtein, some other techniques can be used for this purpose. These are not implemented in Optimus but are mentioned here to give you the best overview of the techniques at hand, as follows:

- **Jaro-Winkler distance**: This distance is a formula consisting of five parameters determined by two compared strings (A, B, m, t, l) and p chosen from [0, 0.25].

- **Damerau-Levenshtein distance**: Damerau-Levenshtein is a modified version that also considers transpositions as single edits.

- **Hamming distance**: Hamming distance is calculated using the number of positions with the same symbol in both strings, provided both strings are of the same length.

- **Q-gram**: A q-gram is the sum of absolute differences between n-gram vectors of both strings.

- **Jaccard index**: Jaccard distance is calculated as 1 minus the quotient of shared n-grams and all observed n-grams.

- **Longest common subsequence (LCS) edit distance**: LCS edit distance is defined as the minimum number of symbols that must be removed in both strings until the resulting substrings are identical.

- **Cosine similarity**: Cosine similarity is considered as 1 minus the cosine similarity of both n-gram vectors.

Now that we have seen all the algorithms available in Optimus to bucket similar strings, let's see how we can use a cluster to get our data into shape.

Applying suggestions

Once we get the clusters from our data, we can start selecting which suggestions we'll be using in our transformed dataset. We can obtain the clusters by running the following code:

```
clusters = df.string_clustering("name", "fingerprint")
```

In the previous example, we're getting `clusters`, which is a custom class with a Python dictionary in it. Let's look at a representation of what we're storing in `clusters`, as follows:

```
print(clusters)
```

This would give us the following output:

```
{
    "name": {
        "johndoe": {
            "suggestions": [
                "John Doe", "john doe", "doe, john", "jóhn dóe"
            ],
            "suggestions_size": 4,
            "total_count": 5
        },
        "alice": {
            "suggestions": ["alice", "álice"],
            "suggestions_size": 2,
            "total_count": 4
        }
    }
}
```

As we can see, here's a dictionary with one element called "name"; if more columns were requested on df.string_clustering, then there'd be more elements in the root of the dictionary. In "name", there's another dictionary; this time, with each cluster of strings is a key to that dictionary, which is a suggestion of the element that would replace all the values in "suggestions". In "total_count", there's a count of the total of elements in the dataframe that are part of the cluster, and on "suggestions_size", the size of the list of suggestions is provided.

Let's say we don't want either of those values; in that case, we must set the suggestions using clusters.set_suggestion, as shown in the following code snippet:

```
clusters.set_suggestion("alice", "Alice")
clusters.set_suggestion("johndoe", "John Doe")
```

We can also set the suggestions by using the index, this being given by the order of appearance of each suggestion in its respective dictionary. Here's the code to accomplish this:

```
clusters.set_suggestion(0, "John Doe")
clusters.set_suggestion(1, "Alice")
```

By default, `set_suggestion` does search the suggestion in the first column of a cluster, as illustrated in the following code snippet. If you want to specify to which column we are setting the values, you can use the third argument:

```
clusters.set_suggestion(0, "John Doe", "name")
clusters.set_suggestion(1, "Alice", "name")
```

Alternatively, we can instead use `set_suggestions` (in plural) to set each value in an ordered array, like this:

```
clusters.set_suggestions(["John Doe", "Alice"])
```

Once we set the values, we'll get the following results on `clusters`:

```
{
    "name": {
        "John Doe": {
            "suggestions": [
                "John Doe", "john doe", "doe, john", "jóhn dóe"
            ],
            "suggestions_size": 4,
            "total_count": 5
        },
        "Alice": {
            "suggestions": ["alice", "álice"],
            "suggestions_size": 2,
            "total_count": 4
        }
    }
}
```

To apply that replacement, we can use `df.cols.replace`, which can use a dictionary with the following format:

```
{
    "name": {
        "John Doe": [
            "John Doe", "john doe", "doe, john", "jóhn dóe"
        ],
```

```
        "Alice": ["alice", "álice"]
    }
}
```

As easily as that, you can clean your data in a faster way instead of manually finding similar entries on a column, which would take a while, if you don't want an outlier to eventually show up.

Let's review every string cluster method available in Optimus. Be sure to select the best method according to the size of the data you're handling. You can see an overview of the methods here:

Method	Speed	Accuracy
Fingerprinting	High	Low
N-gram fingerprinting	High	Low
Metaphone	High	High
Levenshtein	Low	High

Figure 6.4 – String cluster method speed versus accuracy

Remember that processing time can grow higher with the Levenshtein and string grouper methods, so if your data doesn't require that level of accuracy, you can use any other method.

Summary

In this chapter, we learned about the methods we can use in Optimus to group similar string values in a column using key collision and nearest-neighbor methods and replace them with a single value that could represent them better.

With the clustering already created, we learned how to explore suggestions, modified them, and applied them to our data.

Also, we learned about different algorithms that are available in Optimus, which to use depending on the type of data we're handling, and how accurate/fast we need to get our clusters.

In the next chapter, we will learn how to start doing feature engineering to our dataset as an introduction to the **machine learning** (**ML**) chapter.

7
Feature Engineering

Now that we have covered some considerable ground on how to shape our data as needed, let's talk about feature engineering.

If you want to create a machine learning model, you input data. This input data includes the features that an algorithm needs to create a model. These features need to have specific characteristics; for example, it cannot have null values or the data needs to comply and have specific probability distributions.

With featuring engineering, you can prepare the input dataset so that it complies with the algorithm's requirements, and also improve the performance of the machine learning model, thereby creating new features with data we already have.

So, in this chapter, we will be covering the following topics:

- Handling missing values
- Handling outliers
- Binning
- Variable transformation
- One-hot encoding
- Feature splitting
- Scaling

Technical requirements

Optimus can work with multiple backend technologies to process data, including GPUs. For GPUs, Optimus uses **RAPIDS**, which needs an NVIDIA card. For more information about the requirements, please go to the *GPU configuration* section of *Chapter 1*.

You can find all the code for this chapter at `https://github.com/ PacktPublishing/Data-Processing-with-Optimus`.

Handling missing values

One of the most common scenarios when handling data is to find missing values in your dataset.

Missing values are important to handle because, for example, many machine learning algorithms cannot have missing values if you want them to work properly. Or, if you are creating a report, you do not want to present stats with an aggregation of null values.

It's important to notice that Optimus treats `None` and `NaN` (**Not a Number**) values as interchangeable to indicate null values. To handle them, you can do two things: remove the data or impute it. In this section, we will present how Optimus can help with both tasks without providing an exhaustive statistical explanation of when to use each method. Let's see how Optimus can help us with both tasks.

Removing data

In this case, we will see how we can remove whole rows or columns that contain missing values.

Removing a row

First, let's create a dataframe with some null values in many columns:

```
import numpy as np
df = op.create.dataframe({
    "A":[11,2,3,45,6,np.nan,2],
    "B":[1,2,np.nan,45,6,2,3],
    "C":[1,2,3,45,6,2,np.nan],
    "D":[1,2,3,45,6,2,np.nan],
    "E":[1,2,3,45,6,2,np.nan]
})
df.print()
```

A (float64)	B (float64)	C (float64)	D (float64)	E (float64)
11	1	1	1	1
2	2	2	2	2
3	nan	3	3	3
45	45	45	45	45
6	6	6	6	6
nan	2	2	2	2
2	3	nan	nan	nan

To remove all rows with missing values, use the following command:

```
print(df.rows.drop_na())
```

A (float64)	B (float64)	C (float64)	D (float64)	E (float64)
11	1	1	1	1
2	2	2	2	2
45	45	45	45	45
6	6	6	6	6

Removing columns

To see how we can remove columns that contain null values, we must first create a dataframe for it:

```
df = op.create.dataframe({
    "A":[11,2,3,45,6],
    "B":[1,2,None,45,6],
    "C":[1,2,3,45,6]
})
```

Let's print the dataframe to have a better idea of where the null values are:

```
df.print()
```

A (int64)	B (float64)	C (int64)

11	1	1
2	2	2
3	nan	3
45	45	45
6	6	6

Then, delete the B column:

```
print(df.cols.drop("B"))
```

A	C
(int64)	(int64)
---------	---------
11	1
2	2
3	3
45	45
6	6

Remember that you can also pass a list of columns names, like so:

```
print(df.cols.drop(["A", "B"]))
```

In the previous example, we deleted columns A and B. Let's see how else we can handle missing values.

Imputation

Imputation refers to updating missing data with alternate values. The reason you may have missing data could be due to human error while filling in a survey, or even interruptions in a data stream from a flow sensor.

Imputation is the preferable option compared to dropping because you never know how the data you decide to drop can affect the model's performance.

Optimus provides functions that can handle numerical string data. Let's see how it works.

Numerical imputation

Optimus relies on the `impute` method. With this method, you can easily handle continuous values. You can apply one of the following four techniques to handle null values:

- Mean
- Median
- Most frequent
- Constant

Let's see some examples.

First, let's create a dataframe with a column of numbers:

```
import numpy as np
df = op.create.dataframe({"A":[1,2,3,45,6,2,np.nan]})
df.print()
```

This will print a column of integers with the last values represented as `nan`:

```
        A
   (float64)
  -----------
          1
          2
          3
         45
          6
          2
        nan
```

Now we will use `impute` to calculate the `median` value of all the values and apply it to the `nan` value:

```
df.cols.impute("A",data_type="continuous", strategy="mean")
```

Now, replace the `nan` value with `9.83333`, which is the mean value:

```
        A
   (float64)
```

1
2
3
45
6
2
9.83333

With the mean, you can apply the median strategy to replace nan, as follows:

```
df.cols.impute("A",strategy="median").print()
```

You will get the following output:

A
(float64)

1
2
3
45
6
2
2.5

Also, you can use the most frequent value (this is also known as categorical imputation), like so:

```
print(df.cols.impute("A",strategy="most_frequent"))
```

You will get the following output:

A
(float64)

1
2
3
45

```
                6
                2
                2
```

If your data does not match any previous case, you can substitute the nan value with whatever value you input:

```
print(df.cols.impute("A", strategy="constant",
                     fill_value=1))
```

This will print the following output:

```
           A
     (float64)
-----------
           1
           2
           3
          45
           6
           2
           1
```

Now that we know how to handle numeric columns, let's learn how to handle string values.

String imputation

When it comes to string columns, Optimus only gives you the most_frequent method to work with. Let's create a dataframe to work with:

```
df = op.create.dataframe({
    "A":[1,2,3,45,6,2,3],
    "B":["Optimus", "Bumblebee", "Eject", "Optimus",
         "Bumblebee", "Eject", np.nan]
})
```

It works in the same way as it does with string values:

```
print(df.cols.impute("B", strategy="most_frequent"))
    A    B
```

1	Optimus
2	Bumblebee
3	Eject
45	Optimus
6	Bumblebee
2	Eject
3	Bumblebee

As you can see, there are plenty of options to impute a missing value. Now, it's time to learn how to handle outliers.

Handling outliers

An outlier is a data point that is far away and not similar to all the other data points in a sample:

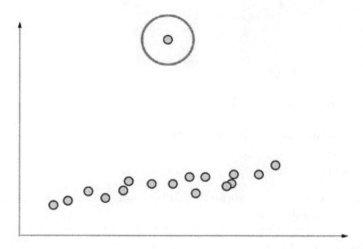

Figure 7.1 – Outlier (marked in red)

Outliers can be detected using graphical (box plots) and not graphical methods, with the graphical methods being more intuitive. Let's talk about the non-graphical statistical methods:

- Tukey or percentiles
- Z-score
- Modified z-score

To show how Optimus can handle outliers, let's create a dataset with positive and negative extrema while considering all the data. Their values will be between 40 and -50:

```
df = op.create.dataframe(
        {"A":[1,2,3,45,6,-50,np.nan],
         "B":["Optimus","Bumblebee","Eject","Optimus",
              "Bumblebee","Eject",np.nan] })
```

Now, let's apply all three methods.

Tukey

Tukey is a mathematical method for detecting outliers. Here, quartiles are used. Optimus uses this method to calculate lower and upper bounds, as follows:

- Lower bound: Q1 - 1.5*IQ
- Upper bound: Q3 + 1.5*IQ

Here, Q1 is the first quartile, Q3 is the third quartile, and IQ is the interquartile range.

In Tukey, every data point that falls outside this range of values is considered an outlier.

Using the `info` method, you can get a general overview of this method:

```
df.outliers.tukey("A").info()
```

In the case of Tukey, you can get a count of the outliers and non-outliers with the upper and lower bound values:

```
{
  'count_outliers': 2,
  'count_non_outliers': 4,
  'lower_bound': -4.75,
  'lower_bound_count': 1,
  'upper_bound': 11.25,
  'upper_bound_count': 1,
  'q1': 1.25,
  'median': 2.5,
  'q3': 5.25,
  'iqr': 4.0
}
```

The `lower_bound` and `upper_bound` values represent the limits after which a value is considered an outlier. In this case, the values that are considered to be outliers are those outside the range of -4.75 and 11.25.

Now, if you just want to get the outlier values, you can use the `select` method:

```
print(df.outliers.tukey("A").select())
```

This will return a dataframe containing the outliers:

A (float64)	B (object)
45	Optimus
-50	Eject

If you want to drop the outliers, you can use the `drop` method:

```
print(df.outliers.tukey("A").drop())
```

As you will see, the value 45 is considered an outlier and was removed from the dataset:

A (float64)	B (object)
1	Optimus
2	Bumblebee
3	Eject
6	Bumblebee
nan	nan

You can also select the values above the upper bound and below the lower bound using the `select_upper_bound()` and `select_lower_bound()` methods, respectively:

```
Print(df.outliers.tukey("A").select_lower_bound())
```

By doing this, you will get the bottom outliers of the dataframe:

A (float64)	B (object)
-50	Eject

If you want to select the data above the upper level bound, you can use `select_upper_bound()`, like this:

```
print(df.outliers.tukey("A").select_upper_bound())
```

By doing this, you will get the top outliers of the dataframe:

```
        A        B
   (float64)  (object)
 -----------  ----------
        45    Optimus
```

Tukey has other helpful methods you can use to get more information about the Tukey method's outcome.

To count the non-outlier values, you can use the `non_outliers_count` method:

```
df.outliers.tukey("A").non_outliers_count()
```

This will print the following output:

```
4
```

To count the outliers, use the following command:

```
df.outliers.tukey("A").count()
```

You will get the following output:

```
3
```

To get information about the quartiles and the whiskers, use the following command:

```
df.outliers.tukey("A").whiskers()
```

The preceding example will print a Python dictionary that contains the following values:

```
{'lower_bound': -4.75, 'upper_bound': 11.25, 'q1': 1.25,
'median': 2.5, 'q3': 5.25, 'iqr': 4.0}
```

Just like Tukey, you can use Z-score to get different bounds.

Z-score

The Z-score is a very useful concept in the field of statistics. It shows us whether a data point is deviating from the mean for a set of values, and if it is deviating, how far away it is. More specifically, the Z-score tells us how much standard deviation a data point has, compared to the mean.

If any data point has a Z-score more than 22, the data point is very different from the rest of the data points, and could be considered as an outlier.

In Optimus, you can indicate how many standard deviations away a point should be from the mean so that it's considered an outlier, by using the `threshold` parameter:

```
threshold=2
print(df.outliers.z_score("A", threshold).select())
```

This will return a dataframe containing the outlier:

A
(float64)

31
-21

To better understand this, let's calculate the Z-score of the A column:

```
print(df.cols.z_score("A"))
```

This will return the following output:

A
(float64)

-0.224362
-0.143592
-0.0628213
2.19875
0.17949
-2.00131
nan
-0.143592
0.017949

If you want to get rid of the outlier, you can use the `drop` method, as follows:

```
print(df.outliers.z_score("A", threshold).drop())
```

This will return the following output:

```
         A
  (float64)
----------
         1
         2
         3
         6
       nan
         2
         4
         6
```

As with the Tukey method, we can get the data above and below the lower and upper bounds, respectively.

If you want to select the data above the upper level bound, you can use `select_lower_bound()`, like this:

```
print(df.outliers.z_score("A").select_lower_bound())
```

This will return the bottom outliers of the dataframe:

```
         A
  (float64)
----------
       -21
```

To select the top ones, you can use the following command:

```
print(df.outliers.z_score("A").select_upper_bound())
```

This will get us the following output:

```
          A
   (float64)
   - - - - - - - - - -
          31
```

Similar to Tukey, in Z_score, you have `info()`, `non_outliers_count()`, and `threshold).count()`.

If you want to avoid misleading boundaries, you can pass a threshold to a modified version of Z-score.

Modified Z-score

We use the Z-score to find out potential outliers, however, this can be inaccurate, especially for smaller sample sizes, because the maximum Z-score is, at most, $(n-1)/sqrt(n)$.

Authors *Iglewicz* and *Hoaglin* recommend using the modified Z-score:

$Mi=0.6745(xi-median(x))/MAD$

Here, **MAD** denotes the **median absolute deviation**.

In Optimus, you can easily apply `modified_z_score`. Let's see how it works.

The aforementioned authors also recommend using a threshold of 3.5:

```
print(df.outliers.modified_z_score("A", threshold=3.5).
select())
```

```
          A
   (float64)
   - - - - - - - - - -
          31
         -21
```

As with the `z_score` method, you can check the modified Z-score values using `modified_z_score`:

```
print(df.outliers.modified_z_score("A", threshold).select())
```

```
          A
   (float64)
```

```
- - - - - - - - - -
       0.6745
      0.33725
            0
        9.443
      1.01175
        8.094
          nan
      0.33725
      0.33725
```

You can drop the outlier with the following command:

```
print(df.outliers.modified_z_score("A", threshold).drop())
```

You will get the following output:

```
         A
    (float64)
- - - - - - - - - -
         1
         2
         3
         6
        nan
         2
         4
         6
```

Similar to the Z-score, with the modified Z-score, you also count with the `info`, `count`, and `non_outliers_count` methods. All three methods group our data to get the outliers, but we can also group this data in a custom way by using binning. Let's take a look.

Binning

The idea of binning is to group some values into a specific category, thus reducing the amount of unique values in a dataset.

For example, let's say we're creating a column of numbers, like so:

```
df = op.create.dataframe(A=[1,2,3,31,6,-21,np.nan,2,4,6])
```

A
(float64)

1
2
3
31
6
-21
nan
2
4

Here, we can group the values in low, medium, and high bins. For this, we can use the cut method:

```
df.cols.cut("A", bins = [0,4,6,35] ,
            labels = ["low", "medium","high"])
```

The cut method will assign the low value to any value between 0 and 4, medium to any value between 4 and 6, and high to any value between 6 and 35:

A
(category)
----
low
low
low
high
medium
nan
nan
low
low

It's important to talk about how bins work. Optimus will include the right-most edge but not the left one. The bins we are using [0, 4, 6, 35] will be represented as (0,4], (4,6], (6, 35]. This means that when binning, Optimus will not take the 0 value but it will take 4 for the first bin, and will not take 4 but will take 6 for the second one:

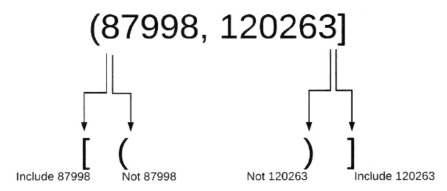

Figure 7.2 – How cut works in Optimus

You can also apply cut to categorical data, as follows:

```
df = op.create.dataframe(A=["Maracaibo", "Caracas", "CDMX",
                    "Monterrey", "Bogota"])
print(df.cols.cut("A", ["Maracaibo", "Caracas", "CDMX",
                "Monterrey", "Bogota"],
            labels=["Venezuela", "Venezuela",
                "Mexico", "Mexico", "Colombia"]))
```

This will return a dataframe in which we group every state to their respective state:

```
A
(object)
----------
Venezuela
Venezuela
Mexico
Mexico
Colombia
```

Binning is commonly used to make a more robust model, thus preventing overfitting at the cost of data loss and performance. It can be helpful for categorical columns to unite values, thus reducing the total amount of unique values.

However, it mainly provides categories for columns with numerical data, sacrificing resolution. This may be redundant for some kinds of machine learning algorithms.

Variable transformation

Some machine learning models, such as linear and logistic regression, assume that the variables follow a normal distribution. More likely, variables in real datasets will follow a more skewed distribution.

By applying several transformations to these variables, and mapping their skewed distribution to a normal distribution, we can increase the performance of our models.

Plotting a histogram or using Q-Q plots could give you an idea of whether the data has a normal distribution or is skewed.

Next, we will look at four methods you can use to adjust your data distribution.

Logarithmic transformation

This is the simplest and most popular among the different types of transformations and involves a substantial transformation that significantly affects the distribution shape.

We can use it (natural logarithmic ln or log base 10) to make extremely skewed distributions less skewed, especially for right-skewed (or positively skewed) distributions.

In Optimus, you can use the `log` method:

```
print(df.cols.log("A"))
```

This will apply the log to the column:

A
(float64)

0
0.30103
0.477121
1.65321

```
         0.778151
         0.845098
              nan
```

This method transforms the values from A into its logarithms. Another method that can be used to transform positively skewed distributions is square root. Let's take a look.

Square root transformation

Another simple transformation, this one has an average effect on the distribution shape: it's weaker than logarithmic transformation, and it's also used to reduce positively skewed distributions.

One advantage of square root transformation is that you can apply it to zero values.

In Optimus, you can use the sqrt method for this:

```
print(df.cols.sqrt("A"))
           A
    (float64)
    -----------
           1
     1.41421
     1.73205
      6.7082
     2.44949
     2.64575
         nan
```

As we already know, this method will return the transformed dataframe, but now with square root values on the A column. As well as square root transformation, you can use reciprocal transformation for right-skewed distributions.

Reciprocal transformation

Reciprocal transformation is a powerful transformation with a radical effect. The reciprocal reverses the order among values of the same sign, so large values become smaller. The negative reciprocal preserves the order among values of the same sign.

You should note that this function is not defined for zero.

In Optimus, you can use the `reciprocal` method like so:

```
print(df.cols.reciprocal("A"))
         A
   (float64)
-----------
         1
       0.5
  0.333333
 0.0222222
  0.166667
  0.142857
       nan
```

As you can see, the values shown here are the inverse of the input values. You can use other known transformations such as exponential or power transformations.

Exponential or power transformation

Power transformation has a reasonable effect on the distribution shape; generally, we apply power transformation (power of two, usually) to reduce left skewness.

In Optimus, you can use the `pow` or `exp` methods for this. Try and see which one gives you better results:

```
print(df.cols.pow("A", 2))
         A
   (float64)
-----------
         1
         4
         9
      2025
        36
        49
       nan
```

As you can see, the values changed dramatically.

But what if we're not using a numerical column? For this, we have a method called `string_to_index`. Let's take a look.

String to index

String to index assigns a numeric value to every identical value in a column. Let's take a look at how it works.

First, let's create a dataframe that contains a couple of repeated values:

```
df = op.create.dataframe({
    "A":["Optimus","Bumblebee","Eject","Optimus","Eject"]
})
```

This will give you a dataframe that looks like this:

```
A
(object)
----------
Optimus
Bumblebee
Eject
Optimus
Eject
```

Now, let's apply the `string_to_index` method:

```
print(df.cols.string_to_index())
```

This will create a column and assign a value of 0 to `Bumblebee`, 1 to `Eject`, and 2 to `Optimus`:

```
A                A_string_to_index
(object)              (int32)
----------      --------------------
Optimus                   2
Bumblebee                 0
Eject                     1
Optimus                   2
Eject                     1
```

In this example, we assigned an index to every value on A. We can do this to allow machine learning algorithms to use this column as a numeric one. Now, let's look at another method that also allows this but in a better way, called one-hot encoding.

One-hot encoding

One-hot encoding is a process where categorical data is converted into an alternate form that is much easier to use for machine learning algorithms, which in turn results in better predictions.

To illustrate how it works, let's say we have the following dataframe:

```
df = op.create.
dataframe({"A":["Optimus","Bumblebee","Eject", "Megatron"],
"B":["Transformer","Transformer","Transformer","Decepticon"]})
```

A	B
(object)	(object)
----------	----------
Optimus	Transformer
Bumblebee	Transformer
Eject	Transformer
Megatron	Decepticon

Most machine learning algorithms can only work with numbers, so, with one-hot encoding, we will create a column containing the category name and assign 0 or 1 to the row if the observation belongs to a specific category:

```
print(df.encoding.one_hot_encoder("B"))
```

This will result in the following output:

A	B	B_Decepticon	B_Transformer
(object)	(object)	(uint8)	(uint8)
----------	----------	---------------	---------------
Optimus	Transformer	0	1
Bumblebee	Transformer	0	1
Eject	Transformer	0	1
Megatron	Decepticon	1	0

Here, Megatron belongs to the Decepticon category, so the number 1 is assigned to the B_Decepticon column, while 0 is assigned to B_Transformer.

However, if our values are more complex than just categories, we need to apply other methods that split these values across multiple columns.

Feature splitting

Feature split is a technique that consists of splitting values from one column to create new ones. A good example could be to split first names and last names that have been saved in a single column into two separate ones, or splitting a date into three columns with separate values for days of the month, months, and years. The main goal of splitting a feature is to give a machine learning algorithim data in small packages that it can interpret better and, by the end, improve the machine learning model's performance.

For featuring splitting, we can use the `unnest` method, which we looked at in Chapter 3. However, there, we focused on how we can produce features to feed our machine learning algorithm.

First, let's start with a dataframe that contains some string values:

```
df = op.create.dataframe({"A":["Argenis Leon","Luis
Aguirre","Favio Vasquez",np.nan]})
print(df.cols.unnest("A"," ", drop=True))
```

The `drop` parameter will delete the column you are splitting, returning the name and last name and any other column in the dataframe:

A_0	A_1
(object)	(object)
----------	----------
Argenis	Leon
Luis	Aguirre
Favio	Vasquez
nan	

Another popular case is to split dates into days, months, and years. First, let's create a dataframe that contains some dates:

```
df = op.create.
dataframe({"A":["10/04/1980","20/05/1995","01/08/1990",np.
nan]})
```

Now, let's split them into three columns while preserving the original column:

```
df.cols.unnest("A", "/", splits=3,
                output_cols=["day","month","year"])
```

A	day	month	year
(object)	(object)	(object)	(object)
----------	----------	----------	----------
10/04/1980	10	04	1980
20/05/1995	20	05	1995
01/08/1990	1	08	1990

As we already know, `unnest` divides our values by a given separator, which is very useful for data preparation for machine learning. Another way to prepare data is by scaling numerical values. Let's take a look.

Scaling

Scaling consists of bringing numerical features in a dataset into the same range of values. For example, in a dataset, you could expect to have an age range between 30 and 75 years and salaries between 30,000 USD and 120,000 USD. Because the scale of both features is very different, this can hurt the model's performance.

Although scaling is not mandatory for many algorithms, some based on distance calculations, such as k-NN or k-means, need to have scaled continuous features to perform well.

To help you with this task, Optimus gives you three scaling methods:

- Normalization
- Standardization
- Max abs scaler

To show you how they work, let's start by creating a simple dataframe:

```
df = op.create.dataframe({"A":[1.12,3.2,4.35,6.3,7.3,np.nan]})
```

Now, let's learn how to apply normalization.

Normalization

Normalization (also called min-max normalization) scales all the values in a fixed range between 0 and 1. In Optimus, you can use the `min_max_scaler` method by using the `cols` accessor, like so:

```
print(df.cols.min_max_scaler("A"))
```

Here, the output will be scaled:

```
         A
  (float64)
-----------
         0
   0.33657
  0.522654
  0.838188
         1
       nan
```

Remember that you can always use the `output` parameter to output the result to another column, like so:

```
print(df.cols.min_max_scaler("A", output_cols="A_normalized"))
```

This is useful for maintaining both columns in the dataset, since the method applies a function that cannot be reverted. Let's look at a similar method called `standard_scaler`.

Standardization

Standardization (or Z-score normalization) rescales the values to ensure that the mean is 0 and the standard deviation is 1. In Optimus, you can use the `standard_scaler` method for this:

```
print(df.cols.standard_scaler("*"))
```

This will result in the following scaled column:

```
         A
  (float64)
-----------
```

-1.51526
-0.569926
-0.0472666
0.838981
1.29347
nan

As we can see, we'll get values over 1 and below 0, unlike normalization. Another type of scaling method is max abs scaler. Let's take a look.

Max abs scaler

This method is used to scale a feature using its maximum absolute value.

This estimator modifies each of the features so that the maximum absolute value for each of them is exactly 1.0, for the training set. It doesn't shift or change the data, and so doesn't remove any of the consistency.

In Optimus, you can use the `max_abs_scaler` method like so:

```
print(df.cols.max_abs_scaler("*"))
```

In return, you will get a column that contains the scaled values:

A (float64)

0.153425
0.438356
0.59589
0.863014
1
nan

As we can see, the maximum value of this result is 1.

We have plenty of options to scale the values of a column for different cases.

Summary

In this chapter, we covered a lot of techniques for preparing our data to be consumed by machine learning algorithms.

One of these techniques is imputation, which is useful for data that contains null values. For data that contains unexpected values, we can apply outlier handling.

By using binning, we can categorize numeric data. If our numeric data is not correctly distributed, we can remove skewness by applying variable transformations, using methods we looked at in the previous chapters.

On the other hand, one-hot encoding allows us to separate the values from a column into multiple Boolean columns. We can split one value that contains lots of data into multiple values by using feature split. Finally, we learned how to scale our data by using multiple methods.

Now that you know about all these techniques, you can make your first steps into machine learning.

In the next chapter, we will learn how to use the data we've prepared so far to create models using the methods available in Optimus.

Section 3: Advanced Features of Optimus

In this section of the book, you will deep dive into advanced applications, including feature engineering, machine learning, and natural language processing functions available in Optimus.

This section comprises the following chapters:

- *Chapter 8, Machine Learning*
- *Chapter 9, Natural Language Processing*
- *Chapter 10, Hacking Optimus*
- *Chapter 11, Optimus as a Web Service*

8
Machine Learning

Up to now, we have covered all the tools needed to create **machine learning** (**ML**) models from data cleaning, data exploration, and feature creation.

Now, we will explore how Optimus can help you easily create, evaluate, and use the most common ML algorithms in a line of code. This way, you won't have to use technologies other than Optimus.

So, in this chapter, we will learn about the following topics:

- Optimus as a cohesive **application programming interface** (**API**)
- Implementing a train-test split procedure
- Training models in Optimus

Technical requirements

Optimus can work with multiple backend technologies to process data, including **graphics processing units** (**GPUs**). For GPUs, Optimus uses the **Real-Time Automated Personnel Identification System** (**RAPIDS**), which needs an NVIDIA card. For more information about the requirements, please go to the *GPU configuration* section in *Chapter 1, Hi Optimus!*.

You can find all the code for this chapter at https://github.com/PacktPublishing/Data-Processing-with-Optimus.

Optimus as a cohesive API

The main goal of Optimus is to create a cohesive API so that you can handle data and create ML models in the simplest way possible. In Optimus, you have the `ml` accessor, which will give you access to the ML algorithms implemented in Optimus.

ML algorithms can be hard to implement in parallel—for example, **density-based spatial clustering of applications with noise (DBSCAN)** is not implemented in Spark. For Optimus, we implemented algorithms that were common to all the libraries, and the ones that we considered as must-haves but that were missed, in a specific library. First, let's see which library empowers every Optimus engine, as follows:

- pandas uses scikit-learn.
- Dask uses Dask-ML.
- cuDF uses cuML.
- Dask cuDF uses cuML.
- Vaex uses vaex.ml.
- Spark uses MLlib.
- Ibis has no ML library available yet.

With this said, now let's see which algorithms are implemented in every library. Have a look at the following overview:

Engine	Library used to support ML	Linear regression	Logistic regression	K-means	Principal
pandas	scikit-learn	x	x	x	x
Dask	Dask-ML		x	x	x
cuDF	cuML		x	x	x
Vaex	vaex.ml		x	x	x
Spark	MLlib		x	x	x

Figure 8.1 – Algorithms that are implemented in every library

As with dataframes, Optimus tries to create a layer to abstract low-level details. Let's enumerate some details so that you can be sure where Optimus can add value, as follows:

- As we saw, every engine has its own library to handle ML, so you do not have to learn to use every library.
- pandas/Dask algorithms rely heavily on NumPy, so you have to handle some low-level data transformation when creating models.

- Spark does not follow the scikit-learn API that follows the rest of the ML libraries. So, if you want to use Spark, you need to learn a new API.

By creating a layer to abstract those details, we'll be providing a simpler interface to Optimus' users. Now, let's see more about how Optimus can help.

How Optimus can help

When you want to create a model, there is a process to ensure that your model can get the best performance it can achieve. After being sure that you have cleaned, prepareed, and feature engineered the data you now need, and depending on the model you want to implement, you will need to split your data into train and test data and evaluate your data using a k-fold method. A general overview of how Optimus processes models internally is expressed in the following diagram:

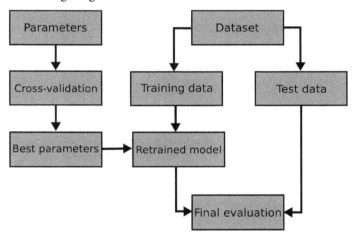

Figure 8.2 – Optimus process to create a model

Let's see how Optimus can help with this task.

Implementing a train-test split procedure

The main idea of splitting your data into two datasets is that you can train your model in one and then test your model performance over new data. When a dataset is split into a training and testing set, the majority of the data goes to the training set and a small part of it is used for testing.

The subset used to fit a model is known as the training dataset. This contains example **inputs and outputs (I/Os)** that will train the model fitting the parameters.

On the other hand, when the inputs on the test dataset are provided to the model, the resulting predictions made from those inputs are then compared to the expected values to assess the model's accuracy.

When to use a train-test split procedure

A train-test split evaluation procedure can be used for classification or regression problems.

The dataset to be used should be large enough to represent the problem domain, covering every common case and enough uncommon cases. Depending on this, it may require thousands or even millions of examples in the whole dataset.

Unless the model has enough data, it won't be able to effectively map the inputs to the outputs, and also wont have the required data in the test set, to generate performance metrics for the model.

Another reason this procedure is handy is to get better computational efficiency. Some of the models can be very expensive when it comes to training, and multiple rounds of evaluation can be hard to deal with.

Test size

A train-test split evaluation has one key parameter, which is the split ratio of our dataset. This is typically represented as a number between 0 and 1 for either the train or the test set, which means that a test set that has a value of 0.2 will have 20% of the content from the original dataset.

There is no perect ratio or percentage for all the cases. We must allocate the split in such a way that all the projects objectives are covered, including the actual cost of training, evaluating and testing the model, and so on.

Common test sizes include test sets with sizes between 0.2 and 0.5.

Now that we know how to implement a train-test split model evaluation procedure, let's look at how we can use it in Optimus.

In Optimus, this process is handled internally by the model, using the `test_size` parameter. Let's suppose we want to apply a linear regression model and use 20% of the data available as test data to test the model performance. We can achieve this with the following code:

```
df.ml.linear_regression(['reclat','reclong'], 'mass(g)', test_size=0.2)
```

In the previous example, we are applying a linear regression model to the `reclat` and `reclong` columns of the dataset.

Repeatable train-test splits

Rows are assigned to the train and test sets randomly, so that the sample is representative.

So, if you want to compare results, you'll want to set a seed for the pseudo-random number generator used when splitting the dataset. This way, the model will be fitted and evaluated using the same set of values from the original dataset.

This can be achieved by setting the `random_state` argument on the methods we'll be using to create our ML models.

Using k-fold cross-validation

Cross-validation or k-fold cross-validation is a commonly used technique used to evaluate ML models. This consists of splitting the data into k groups and comparing every one of the n parts with the rest. This method is easy to understand and generally results in fewer biased models compared with a train-split method. The method is better represented as follows:

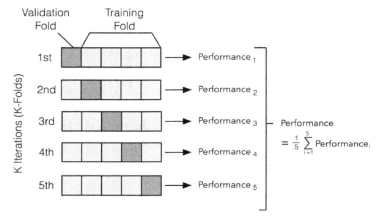

Figure 8.3 – How k-fold cross-validation works

For example, if we test our model on five folds, we will process the data as shown in *Figure 8.3*.

We'll review all these methods to get a better understanding of how to test our models, but internally Optimus uses a k-fold technique.

Training models in Optimus

Now that we know how the test/train, split, and cross-validation processes work, let me tell you something amazing. You don't have to struggle with configuring and writing code to make this process work, as Optimus will do the heavy lifting for you.

Let's see the ML models available in Optimus.

Linear regression

Linear regression is a supervised ML algorithm that is useful for finding out how variables are linked to each other. By assigning a linear equation to the data that we have, we can use fresh data and predict the output, as illustrated in the following diagram:

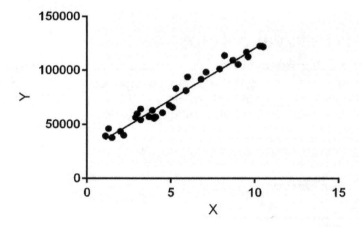

Figure 8.4 – A line approximated to a cluster of points

In the preceding diagram, we can see a line that approximates a cluster of points. Let's see how to calculate this approximation.

First, let's start by creating a dataset with the following code:

```
import numpy as np
size = 10000
```

```
data = {
    'length':[round(random.uniform(1,2),1) for i in
range(size)],
    'width': [round(random.uniform(1,1.5),1) for i in
range(size)],
    'height': [random.randint(20,50) for i in range(size)],
    'type': [random.randint(0,1) for i in range(size)]
}

df = op.create.dataframe(data).repartition(4).execute()
df['weight'] = df['height'] * df['width'] * df['length'] *
[random.uniform(1,1) for i in range(size)]
df = df.cols.round('weight',1)
```

This will create the following output:

length (float64)	width (float64)	height (int64)	type (int64)	d_weight (float64)
1.4	1.2	36	0	60.5
1.1	1.4	20	1	30.8
1.9	1.1	20	0	41.8
1.2	1	30	1	36
1.6	1.1	20	1	35.2
1.8	1.4	39	0	98.3
1.6	1.4	20	0	44.8
1.4	1.4	21	0	41.2
1.5	1.1	38	0	62.7

Now, you can predict a value using a linear regression model, like so:

```
lm = df.ml.linear_regression('height','d_weight', test_
size=0.2,fit_intercept=False)
```

Also, you can predict the d_weight value using the height, like so:

```
lm.predict(36)
```

This will return the following output:

```
[68.18919244183671]
```

In this case, for a height of 36, the model will return a dimensional weight of 68.18919244183671, but let's explore a little what happened under the hood.

df.ml.linear_regression will return an Optimus model object. If you execute lm, it will return the following output:

```
<optimus.engines.pandas.ml.models.Model at 0x1f931014f48>
```

This object has some additional functions that can help you to evaluate and visualize your model.

To evaluate your model or how good your model is at predicting a value, Optimus uses the **R-squared (R^2)** coefficient of determination. This indicates how well the regression predictions approximate the real data points. In Optimus, you can check the accuracy like so:

```
print(lm.evaluate())
```

This will return the following output:

```
{'accuracy': 0.5518192755546748, 'standard deviation':
0.054855397528967634}
```

It's hard to say whether it's now a *good* R^2 value. The threshold for a good R^2 value depends widely on the domain, therefore it's most useful as a tool for comparing different models.

If you want to know where the line intercepts with the axis, you can get this information using the following line of code:

```
lm.intercept()
```

This will give you the following output:

```
-1.0359347377853823
```

Or, if you want to get linear equation coefficients, you can use the following code:

```
lm.coef()
[1.9221350555037229]
```

OK—now that we know how to predict and evaluate our model, let's go back a little bit and see what happened inside df.ml.linear_regression.

Optimus makes two important steps here related to what we learned in the past section, as follows:

- Train-test split
- K-fold cross-validation

Optimus first separates the data into train and test data. Optimus uses by default 20% of the data for testing. You can set how much data you want to use for this with the test_ size property, like so:

```
lm = df.ml.linear_regression('height','d_weight', test_
size=0.3)
```

In k-fold cross-validation, Optimus takes five folds by default. It's important to note that the accuracy we calculated previously is the mean of all the R^2 calculations in every fold. If you want to know how to perform every fold, you can use lm.scores to get the following output:

```
{'neg_mean_absolute_error': [-13.811269861612953,
-11.279855517497198, -12.091910783179417, -12.318287462375567,
-12.317722455904498], 'neg_mean_squared_error':
[-303.27443869789056, -219.26820300534945, -217.3260761387018,
-243.19541612297002, -238.7205727812638], 'neg_root_mean_
squared_error': [-17.41477644696855, -14.807707554018936,
-14.741983453345137, -15.594723983545526, -15.450584868582283],
'r2': [0.4539669486939144, 0.6245671700067864,
0.560437200131502, 0.5421392626142632, 0.5600087733488277]}
```

To finish, let's plot a line produced for the linear regression algorithm that better approximates all data points. For that, we will use the `plot` method. You can see a representation of this here:

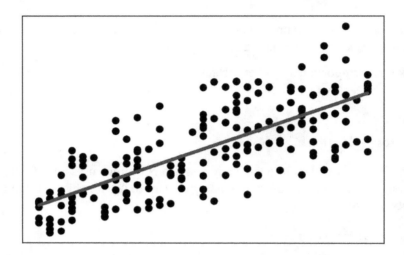

Figure 8.5 – Multiple linear regression

To make predictions using linear regressions, you can implement multiple features.

Let's say we want to predict the d_weight value using the `length`, `width`, and `height` properties. We can do this in the following way:

```
lm = df.ml.linear_regression(['length','width','height'], 'd_
weight', test_size=0.2,fit_intercept=False,)
```

And to predict, you can use the following line of code:

```
print(lm.predict([[1.4,1.38,25]]))
```

Because you are using three independent variables, you are going to get three coefficients. To print these, run the following line of code:

```
print(lm.coef())
```

You will get the following result:

```
[21.26987554210074, -10.097655140541706, 1.364351495308436]
```

And to evaluate the model, as we saw, we can use the following code:

```
print(lm.evaluate())
{'accuracy': 0.9684722231977312, 'standard deviation':
0.0017256479289964421}
```

As we can see, we improve the model accuracy by using more data features.

Logistic regression

Logistic regression is an ML algorithm that can predict the probability of certain discrete values using continuous values. Let's see how this works, using the iris dataset from scikit-learn. The following example consists of 50 samples from each of three species of **iris** plants:

```
import numpy as np
import pandas as pd
from sklearn.datasets import load_iris

iris = load_iris()
df = pd.DataFrame(data=np.c_[iris['data'], iris['target']],
                  columns=iris['feature_names'] + ['target'])
df = op.create.dataframe(df)
```

We will get a larger dataset, so let's print it in two parts, as follows:

```
df.cols.select(['sepal length (cm)', 'sepal width (cm)',
'target']).print()
```

This will show the following output:

sepal length (cm) (float64)	sepal width (cm) (float64)	target (float64)
5.1	3.5	0
4.9	3	0
4.7	3.2	0
4.6	3.1	0
5	3.6	0

And here's the other part:

```
df.cols.select(['sepal length (cm)', 'sepal width (cm)',
'target']).print()
```

This will show the following output:

petal length (cm) (float64)	petal width (cm) (float64)	target (float64)
1.4	0.2	0
1.4	0.2	0
1.3	0.2	0
1.5	0.2	0
1.4	0.2	0

The flower species are represented as 0, 1, and 2, which correspond to the setosa, versicolor, and virginica species.

You can get the names using iris.target_names, resulting in the following output:

```
array(['setosa', 'versicolor', 'virginica'], dtype='<U10')
```

Now, to calculate a logistic regression model using a 20% test size, run the following code:

```
lr = df.ml.logistic_regression([0,1,2,3],'target', test_
size=0.2)
```

This will return the model. Then, you can use predict to return the species taking the features you input, as follows:

```
lr.predict([[5.1,3.5,1.4,0.2]])
```

This will return the following output:

```
[0.0]
```

The preceding output relates to the setosa species.

Now, if you want to know the probability of each species taking the features you input, you can use the following line of code:

```
lr.predict_proba([[5.1,3.5,1.4,0.2]])
```

This will return an array with the probability of every species, as follows:

```
[[8.76409999e-01, 1.23555395e-01, 3.46054257e-05]]
```

To get the accuracy of the model, you can use the following line of code:

```
lr.evaluate()
```

This will result in the following output:

```
{'accuracy': 0.94, 'standard deviation': 0.9}
```

As we can see, we have plenty of options to use logistic regression. Now, we'll show we can get even more detailed information from our classification models by generating some plots.

Model performance

The model performance refers to how well your model can predict future data. Let's see how we can apply some evaluation methods depending on the ML algorithm we use.

Confusion matrix

A confusion matrix is a table that can show the performance of a algorithm on a set of test data. In this section, we will show how to use Optimus to make confusion matrices more user-friendly and easy to understand.

To get a confusion matrix, you can simply call the following method in the model:

```
lr.confusion_matrix()
```

This will result in the following confusion matrix plot:

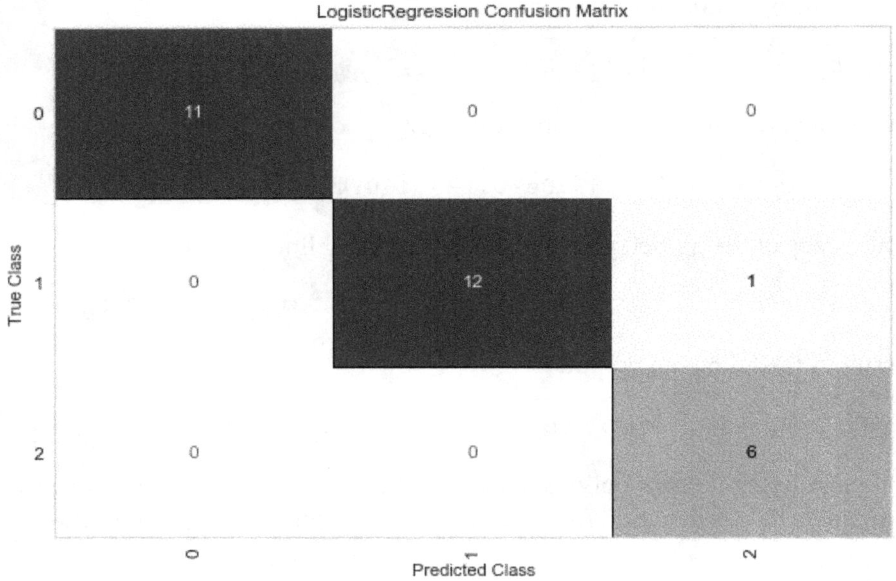

Figure 8.6 – Confusion matrix

What we are seeing here is this:

- The true class of 11 items is 0. None were classified wrongly.
- The true class of 11 items is 12, but one was wrongly classified as 2.
- The true class of six items is 6. None were classified wrongly.

As you can see, by using a confusion matrix, we can have an idea of how accurate a classifier is. Let's see how else we can measure the performance of our model.

ROC

An **AUC-ROC** (**area under the curve** and **receiver operating characteristic**, respectively) curve helps us visualize how well our ML classifier is performing. Specifically, it tells how much a model is capable of distinguishing between different classes. A higher AUC means a more accurate prediction.

A higher *x*-axis value (near the right part of the plot) tells us there is a higher number of false positives, while a higher *y*-axis value (near the top part of the plot) indicates a higher number of true positives. To plot an AUC-ROC curve, run the following line of code:

```
lr.roc_auc()
```

This results in the following output:

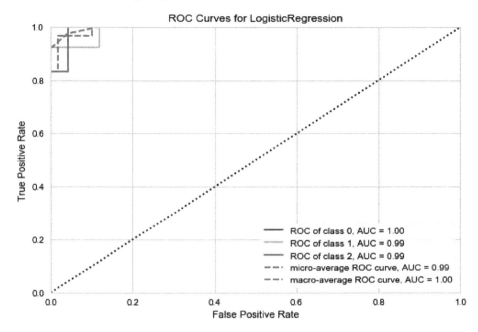

Figure 8.7 – ROC curve plot

Now, let's learn about another measurement technique called **precision-recall** (**PR**).

PR curve plot

To understand the concept of precision and recall, we first need to understand the four types of results we can get, so we can predict a value for this, as follows:

- **True positive**: The prediction is correct and the actual value is positive.

- **False positive**: The prediction is wrong and the actual value is positive.

- **True negative**: The prediction is correct and the actual value is negative.

- **False negatives**: The prediction is wrong and the actual value is negative.

But what does this mean?

Suppose we have a camera system that can identify whether an object is a human or a robot. These are the four possible outcomes:

Positive		Actually is	
		Negative	
System predicts	Positive	It's a robot/It's actually a robot (True positives)	It's a robot/Actually, it's not a robot (False positives)
	Negative	It's not a robot/It's actually a robot (False negatives)	It's not a robot/Actually, it's not a robot (True negatives)

Figure 8.8 – Types of results in an ML model

The precision and the recall are metrics that can measure the performance of a model. The precision is the ratio of actually true positives—this represents how precise a model is in guessing which elements are true. On the other hand, recall is the ratio of relevant elements that are selected—it's calculated by dividing the count of true positives by the sum of the count of true positives and the count of false negatives.

In Optimus, you can plot the precision and recall from a model using the following line of code:

```
lr.precision_recall()
```

This will result in a plot like this:

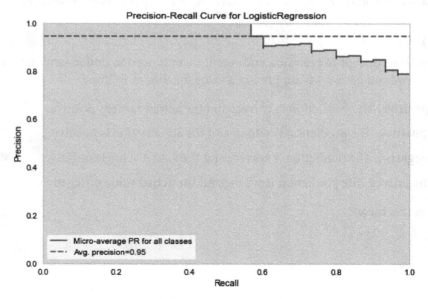

Figure 8.9 – PR curve plot

As you can see, by plotting a PR curve, you'll be able to know the trade-off between the precision and the recall of your model.

K-means

K-means clustering is an **unsupervised learning** (UL) technique used when we have unlabeled data, which means that you do not know which categories or groups every observation belongs to. Optimus requires you to define the number of clusters in which it will iteratively try to put every data point inside a group. In Optimus, you can apply k-means clustering using `ml.k_means`, like so:

```
km = df.ml.k_means([0,1,2,3],'target',3)
km.predict([[5.1,3.5,1.4,0.2],[6.2,2.9,4.3,1.3]])
[0, 1]
km.plot_clusters()
```

The centroids are in **X** and every color represents one of the three clusters, as illustrated in the following screenshot:

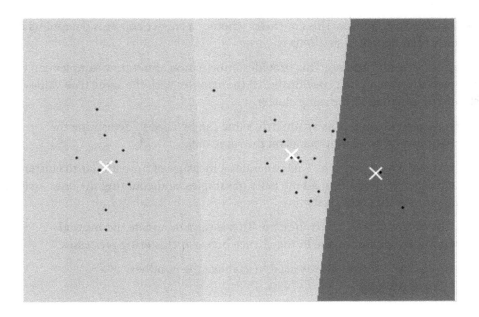

Figure 8.10 – K-means plot

Model evaluation

In Optimus, you can get the k-means evaluation score just by using the `scores` method, like so:

```
km.scores()
{
    'inertia': 139.82049635974974,
    'homogeneity_score': 0.6591265018049008,
    'completeness_score': 0.6598476779627759,
    'v_measure_score': 0.659486892724918,
    'adjusted_rand_score': 0.6201351808870379,
    'adjusted_mutual_info_score': 0.6552228479234864,
    'silhouette_score': 0.5061527484935536
}
```

Let's see what every value means, as follows:

- `inertia`: Inertia is a measure of how consisten clusters are.
- `homogeneity_score`:This considers whether a cluster contains datapoints that are only members of a single class.
- `completeness_score`: This provides information about how samples can be assigned to cluster, more specifically, all the samples with the same true values should be assigned to the same cluster.
- `v_measure_score`: This symmetric value can be used to determine the compatibility of two assignments on the same data.
- `adjusted_rand_score`: This generates a measure of how similar to clusters are to each other, by looking at all the pairs of samples, and counting the ones assigned to the same or different clusters.
- `adjusted_mutual_info_score`: This is used to update the **mutual information** (**MI**) score due to the chance between clustering processes.
- `silhouette_score`: This is used to evaluate the goodness of a clustering technique.

Not knowing the number of centers

In some cases, we have no idea of how many clusters our dataset is divided into. For this, we can use the elbow method.

It's called the elbow method because the chart resembles an arm with an 'elbow' (the point of inflection on the curve), which can be used as an indication of how many clusters exist in our dataset.

The elbow method runs k-means clustering on the dataset k times. For each iteration, the distortion score (or sum of squared errors) is calculated. The idea is that this score could first decrease and then flatten, forming an elbow shape. To create an elbow plot, we can use the following line of code:

```
km.plot_elbow(4, 11)
```

This will output a graph between 4 and 11 clusters, as illustrated in the following screenshot:

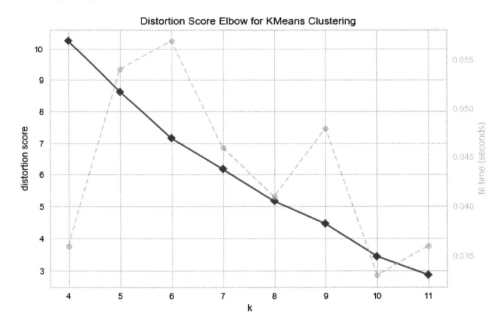

Figure 8.11 – Elbow method plot between 4 and 11 clusters

If you want to plot a graph between two and eight clusters, you can use the following line of code:

```
km.plot_elbow(2,8)
```

This results in the following plot:

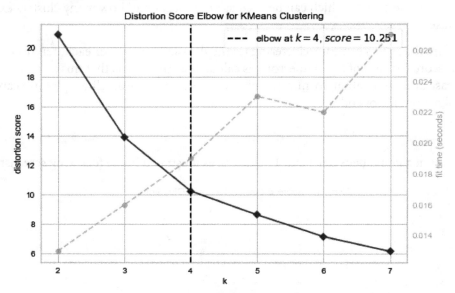

Figure 8.12 – Elbow method plot between two and eight clusters

In the last plot, we see a dashed line. This line defines the elbow of the curve, and it can be used as the number of clusters in our dataset. To be clear, this is a heuristic method, so it can be taken more as guidance than a definite method to calculate the number of clusters.

In *Figure 8.11*, a dashed line was not drawn because an elbow could not be identified.

PCA

For human beings, it is difficult to process information beyond three dimensions. We can use some artifacts such as coloring the data dots or assigning some shape and size to add extra dimensions, but when we are handling hundreds or even dozens of dimensions, it's difficult (if not impossible) to imagine how to visualize the dataset.

That is when PCA comes to our aid. PCA is a dimensionality-reduction method in which you can reduce n dimensions to smaller numbers but still conserve the information from the larger dataset. Let's take a look at how we can reduce n features/columns to two, as follows:

```
df.ml.PCA([0,1,2,3], n_components=2).print(5)
      PCA_0           PCA_1
    (float64)       (float64)
 -----------     -----------
   -2.2647          0.480027
```

```
-2.08096       -0.674134
-2.36423       -0.341908
-2.29938       -0.597395
-2.38984        0.646835
```

Also, you can add any other column from the original dataframe. In this case, let's add the `target` column and use it to plot the color of every species so that we can easily differentiate the three clusters, as follows:

```
print(df.ml.PCA([0,1,2,3], 'target', n_components=2))
```

PCA_0	PCA_1	target
(float64)	(float64)	(float64)
-----------	-----------	-----------
-2.2647	0.480027	0
-2.08096	-0.674134	0
-2.36423	-0.341908	0
-2.29938	-0.597395	0
-2.38984	0.646835	0

Then, we can plot the data, like so:

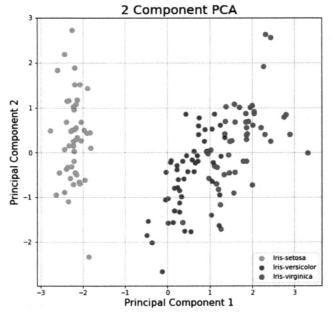

Figure 8.13 – PCA plot

At this point, we have seen how to create models using Optimus. Now, let's see how we can save them to use them whenever we need to.

Loading and saving models

An important point after creating our model is to save it for further use in the future. Happily, with Optimus, you can simply use the `save` method in the model. Let's use our previously created logistic regression model as an example, as follows:

```
lr.save('model.sav')
```

This will save a file named `model.sav`, in which our model is saved.

Now, to load our model back, we use the Optimus object we initially create, like so:

```
m = op.load.model('model.sav')
```

If you want to check which model has been loaded, simply call the m variable, which will print the following information:

```
LogisticRegression(n_jobs=1, solver='liblinear')
```

Now, to predict a value using some features, run the following line of code:

```
m.predict([[6.3,3.3,4.7,1.6]])
```

This will return 2 as the predicted species, as we can see here:

```
array([2.])
```

As you see, you can load and save models in Optimus with a single line of code. This will help you to use your model in the future from a Jupyter notebook to a web server, to be consumed remotely.

Summary

In this chapter, we learned about the available algorithms for each library in Optimus to handle ML. We saw about train-test split evaluation and when to use it.

We also learned about different training models such as linear regression, logistic regression, k-means, random forest, and PCA.

Finally, we learned how to load and save those models for further use or deployment.

In the next chapter, we will learn about how to use the natural language function available in Optimus.

9
Natural Language Processing

Terabytes of text data are created on a daily basis by users of all sorts of software, from enterprise systems to social networks. All this unprocessed data hides amazing opportunities to improve how businesses work.

In this chapter, we will learn how to clean and process our data in order to prepare it to create features that can be used as input to create machine learning models.

The topics we will be covering in this chapter are as follows:

- Natural language processing
- Removing unwanted strings
- Stemming and lemmatization
- `word_tokenizer`
- Feature extraction from text

Technical requirements

Optimus can work with multiple backend technologies to process data, including GPUs. For GPUs, Optimus uses **RAPIDS**, which needs an NVIDIA card. For more information about the requirements, please go to the *GPU configuration* section in *Chapter 1, Hi Optimus!.*

You can find all the code for this chapter at `https://github.com/ PacktPublishing/Data-Processing-with-Optimus`.

Natural language processing

Natural language processing (NLP) is an interdisciplinary sector that lies between linguistics and computer science, primarily aligning with artificial intelligence. While NLP deals with two types of data – text and audio – Optimus is more focused on text data.

Unstructured data, which mostly comes from the web (for example, tweets or Facebook comments), requires preprocessing before it can be further analyzed. Optimus can help you here, as it has the main functions you will need to finish your work. Let's explore how we can remove unwanted strings first.

Removing unwanted strings

To perform NLP, we need to ensure the data doesn't have any special characters, diacritics, HTML tags, or any other content that may make it difficult for the algorithms to work properly.

To see what Optimus can do for you, let's load a dataframe with text to help us demonstrate Optimus' features:

```
df = op.load.csv("text.csv", sep=";")
```

That will load a text we specially prepared from Wikipedia. Let's see its content:

```
text

(object)

"<a>http://google.com</a> <span>Transformers is an American
and Japanese media franchise produced by the American toy
company Hasbro and Japanese toy company Takara Tomy. It follows
the battles of sentient, living autonomous robots, often the
Autobots and the Decepticons, who can transform into other
forms, such as vehicles and animals. The franchise encompasses
toys, animation, comic books,

...
```

Now that we know the data, let's start transforming it.

Stripping the HTML

When getting data from websites, it is common to find HTML tags. Because HTML tags are a way to show information in an internet browser, they rarely add any value to most common NLP tasks, so we tend to remove them.

To use HTML tags, you surround some text with any of the tags available. Some of the tags commonly found are "div," "a," and "spam." In this case, let's see how to remove all HTML tags:

```
print(df.cols.strip_html("text"))
```

This will print the following output:

```
text
(object)
"http://google.com Transformers is an American and Japanese
media franchise produced by the American toy company Hasbro
and Japanese toy company Takara Tomy. It follows the battles
of sentient, living autonomous robots, often the Autobots
and the Decepticons, who can transform into other forms, such
as vehicles and animals. The franchise encompasses toys,
animation, comic books,
...
```

As you can see, the HTML tags <a> and </a> have been removed. This will give us cleaner values. Let's see how we can transform our strings even more.

Removing stopwords

Other than URLs, HTML tags, and special characters, there are other words that are not essential for sentiment analysis or text classification. Words such as I, me, he, you, and so on just increase the size of the text without affecting results, and thus, we should get rid of them.

For our task, we will use a premade collection of stopwords from NTLK, or any NLP library. Alternatively, we can create our collection depending on the task.

By default, Optimus will remove the stopwords in English. For this, you can use the following code:

```
print(df.cols.remove_stopwords("name"))
```

This will remove the stopwords in the text. For example, you can see that words such as "is," "an," "and," and "by" are removed in the following code:

```
text
(object)
"<>http://google.com</> <span>transformers american japanese
media franchise produced american toy company hasbro japanese
toy company takara tomy. Follows battles sentient, living
autonomous robots, often autobots decepticons, transform forms,
vehicles animals. Franchise encompasses toys, animation, comic
books
...
```

If you want to use another language, you can use the language argument.
Let's see an example passing "span"sh":

```
df.cols.remove_stopwor"s("t"xt", langua"e="span"sh")
```

Optimus relies on the **Natural Language Toolkit (NLTK)** to get the stopword list. NLTK, as stated on its website, *is a leading platform for building Python programs to work with human language data.*

To get the list of languages available, you can execute the following commands:

```
from nltk.corpus import stopwords
print(stopwords.fileids())
```

This will print the following output:

```
['arabic', 'azerbaijani', 'danish', 'dutch', 'english',
'finnish', 'french', 'german', 'greek', 'hungarian',
'indonesian', 'italian', 'kazakh', 'nepali', 'norwegian',
'portuguese', 'romanian', 'russian', 'slovene', 'spanish',
'swedish', 'tajik', 'turkish']
```

If you want to get the list of stopwords in a specific language, you can use the following command:

```
print(stopwords.words("english"))
```

This will give us a list of every stopword in English, as shown here:

```
['i', 'me', 'my', 'myself', 'we', 'our', 'ours', 'ourselves',
'you', "you're", "you've", "you'll", "you'd", 'your', 'yours',
'yourself', 'yourselves', 'he', 'him', 'his', 'himself',
'she', "she's", 'her', 'hers', 'herself', 'it', "it's", 'its',
'itself', 'they', 'them', 'their', 'theirs', 'themselves',
'what', 'which', 'who', 'whom', 'this', ...
```

We will also convert all the text to lower case as text in Python is case-sensitive.

Stopwords are not the only thing that may pollute our data. URLs should also be removed from our data in most cases. Let's see how.

Removing URLs

As with HTML tags, it is almost certain that you will find **Uniform Resource Locators (URLs)** when getting data from web pages. A URL is just a piece of text that points to a location on the web.

Because we almost always want to get insight from the text and not from URLs, we can remove them in Optimus using the `remove_urls` function, as shown here:

```
print(df.cols.remove_urls())
```

Remember that if you don't use the column name, you will apply the function to the whole dataframe:

```
text
(object)
"<a></a><span>Transformers is an American and Japanese media
franchise produced by the American toy company Hasbro and
Japanese toy company Takara Tomy. It follows the battles of
sentient, living autonomous robots, often the Autobots and
the Decepticons, who can transform into other forms, such
as vehicles and animals. The franchise encompasses toys,
animation, comic books

...
```

As you can see, this will remove the substring `http://google.com` from the text.

After removing URLs, we can remove other characters. Let's see how.

Removing special characters

Special characters such as `.` (dots) and `;` (semicolons) typically don't add any value, so we can remove them.

Optimus uses the symbols from the Python code library, which we will import using the following command:

```
import string
string.punctuation
```

That will print the following output:

```
'!"#$%&\'()*+,-./:;<=>?@[\\]^_'{|}~'
```

As you can see, between these symbols are symbols such as `$` and `%`. These may be important to preserve to denote money or percentages in our text, which could be useful when dealing with financial data or statistics.

To remove the special characters, you only need the following command:

```
print(df.cols.remove_special_chars("text"))
```

This will print the following output:

```
text
(object)
ahttpgooglecoma spanTransformers is an American and Japanese
media franchise produced by American toy company Hasbro and
Japanese toy company Takara Tomy It follows the battles of
sentient living autonomous robots often the Autobots and the
Decepticons who can transform into other forms such as vehicles
and animals The franchise encompasses toys animation comic
books

...
```

If for some reason you want to remove specific characters, then you can always rely on the `replace` function. To remove hyphens, you can use this command:

```
df.cols.replace("text", "-", "", "chars")
```

Or you can use a list to remove many characters, as seen here:

```
df.cols.replace("text", ["-", "*"], "", "chars")
```

That will replace – and * with a white space.

Now that we have removed useless text from our values, let's homogenize our text by expanding contracted words.

Expanding contracted words

In our everyday verbal and written communication, a lot of us tend to contract common words. One example is how "you are" becomes "you're." Converting contractions into their natural form will help us gain further insight.

First, let's create another dataset for this example:

```
df2 = op.create.dataframe({"text": ["I'll let you know when, it
shouldn't take long. Don't rush."]})
```

Then, we can call `expand_contrated_words` to expand our contracted words, as shown here:

```
df2.cols.expand_contrated_words("text")
```

This will print the following output:

```
text
(object)
I will let you know when, it should not take long. Do not rush.
```

As you can see, the words "I'll," "shouldn't," and "Don't" were replaced by "I will," "should not," and "do not," respectively.

Once we expand the contracted words, we'll need to reduce words to minimal expressions using stemming and lemmatization.

Stemming and lemmatization

In any text, it is common to find a word in multiple forms. See these, for example:

- Truck
- Trucks
- Truck's
- Trucks'

All these words have the unique root `Truck`. The words in the list are called inflections.

The following is a quote from Wikipedia:

> *In grammar, inflection is the modification of a word to express different grammatical categories such as tense, case, voice, aspect, person, number, gender, and mood. An inflection expresses one or more grammatical categories with a prefix, suffix, or infix, or another internal modification such as a vowel change.*

Changing a word from its inflected form to its root form is called **word normalization**.

In natural language processing, there are two main techniques to achieve this: **stemming** and **lemmatization**.

Stemming

While stemming, we use an algorithm to reduce the word to its stems. This is not the case for lemmatization, in which we use the language's morphological root.

There are many algorithms to get the stem of a word. Optimus supports the Porter, Lancaster, and Snowball algorithms. Let's see how it works.

To apply stemming to a column in Optimus, you can use the following command:

```
df.cols.stem_verbs("text")
```

This will return all the columns with the stemmed verbs using Porter (this is the default param value):

```
text

(object)

transform american japanes media franchis produc american toy
compani hasbro japanes toy compani takara tomi follow battl
sentient live autonom robot often autobot decepticon transform
form vehicl anim franchis encompass toy anim comic book video
game film franchis began 1984 transform toy line compris
transform mecha toy takara diaclon microman toylin rebrand
western markets1 term gener 1 cover anim televis seri transform
comic book seri name divid japanes british canadian spinoff
respect sequel follow gener 2 comic book beast war Summary

. . .
```

An improved stemmer method is the `snowball` stemmer. Unlike the `porter` or `lancaster` stemmer, it supports many languages.

You can use the `stemmer` parameter to change the stemmer you want to use. For example, to apply the `snowball` stemmer, use this:

```
print(df.cols.stem_verbs("text", stemmer="snowball",
language="english"))
```

That will return the following output:

```
text
(object)
transform american japanes media franchis produc american
toy compani hasbro japanes toy compani takara tomi follow
battl sentient live autonom robot often autobot decepticon
transform form vehicl anim franchis encompass toy anim comic
book video game film franchis began 1984 transform toy line
compris transform mecha toy takara diaclon microman toylin
rebrand western markets1 term generat 1 cover anim televis seri
transform comic book seri name divid japanes british
...
```

The good thing about Snowball is that it supports multiple languages. As the Snowball functionality is built using `nltk`, you can get the full list of languages supported by importing `nltk`:

```
from nltk.stem import SnowballStemmer
print(", ".join(SnowballStemmer.languages))
```

We'll get the following result:

```
arabic, danish, dutch, english, finish, french, german,
hungarian, italian, norwegian, portuguese, romanian, russian,
spanish, swedish
```

Let's now explore a little about the difference between the three stemmers available in Optimus.

Overstemming and understemming

To get a clearer view of the features of the different stemmer algorithms, let's explore them in three dimensions: languages supported, speed, and aggressiveness, as shown in the following figure:

Stemmer	Multi Language	Speed	Aggressiveness
Porter	English	Slower	Less aggressive
Snowball (Porter2)	15 languages	Faster than Porter	Less aggressive
Lancaster	English	Faster	More aggressive

Figure 9.1 – Stemmer features

In general, Snowball is most appropriate when you need to stem, because it is less aggressive, faster, and is available in many languages. In the words of its creator, Martin Porter, it is an improved version of the Porter stemmer.

Lemmatization

Lemmatization is the process of reducing a word to its lemma or canonical form. For example, "Playing," "Plays," and "Played" would be reduced to "Play."

The reduced form is the linguistic root word. To differentiate this from the stemming process, remember that the word was reduced using some kind of algorithm.

In Optimus, you can call `lemmatize_verbs`, as shown in the following code:

```
df.cols.lemmatize_verbs("text")
```

This will give the following output:

```
text
(object)
transformer american japanese medium franchise produce
american toy company hasbro japanese toy company takara tomy
follow battle sentient living autonomous robot often autobot
decepticon transform form vehicle animal franchise encompass
toy animation comic book video game film franchise begin
1984 transformer toy line comprise transform mecha toy takara
diaclone microman toyline
. . .
```

As we can see, the words "transformers," "produced," and "follows" were changed to "transformer," "produce," and "follow," respectively.

Now that we have reduced every word to its minimal expression, let's tokenize every word.

word_tokenizer

Word tokenization is the process of splitting a large sample of text into words. This is a requirement in NLP tasks where each word needs to be captured and subjected to further analysis, such as classifying and analyzing them for a particular sentiment.

In Optimus, you just need to call `word_tokenizer` in the `cols` accessor, as in the following code:

```
print(df.cols.word_tokenize("text","tokens")["text"])
```

You will then obtain a list of words for every row, as shown here:

```
text
(object)
['transformers', 'american', 'japanese', 'media', 'franchise',
'produced', 'american', 'toy', 'company', 'hasbro', 'japanese',
'toy', 'company', 'takara', 'tomy', 'follows', 'battles',
'sentient', 'living', 'autonomous', 'robots', 'often',
'autobots', 'decepticons', 'transform', 'forms', 'vehicles',
'animals', 'franchise', 'encompasses', 'toys', 'starting',
'2019', 'incarnations', 'story', 'based', 'different', 'toy',
...
```

Now that we have every word in a list, let's explore how we can get extra information for every word.

Part-of-speech tagging

This is the process in which you tag a word depending on its role as a part of speech. The different parts of speech that there are include nouns, pronouns, verbs, adjectives, adverbs, prepositions, conjunctions, interjections, and others.

Each word in a sentence is a part of speech. Tagging those words with this information is what is known as part-of-speech tagging. This includes nouns, pronouns, verbs, adjectives, adverbs, prepositions, conjunctions, interjections, and sub-categories of those parts.

Optimus will return a Python tuple adding one of the following constants depending on the type of word detected. Optimus uses `nltk`, so the possible result for every word is as follows:

- `CC`: Coordinating conjunction
- `CD`: Cardinal digit
- `DT`: Determiner
- `EX`: Existential there (as in "there is"; think of it as "there exists")
- `FW`: Foreign word
- `IN`: Preposition/subordinating conjunction
- `JJ`: Adjective (for example, "big")
- `JJR`: Adjective, comparative (for example, "bigger")
- `JJS`: Adjective, superlative (for example, "biggest")
- `LS`: List marker (for example, "1")
- `MD`: Modal (for example, "could" or "will")
- `NN`: Noun, singular (for example, "desk")
- `NNS`: Noun plural (for example, "desks")
- `NNP`: Proper noun, singular (for example, "Harrison")
- `NNPS`: Proper noun, plural (for example, "Americans")
- `PDT`: Predeterminer (for example, "all the kids")
- `POS`: Possessive ending (for example, "parent's")
- `PRP`: Personal pronoun (for example, "I," "he," or "she")
- `PRP$`: Possessive pronoun (for example, "my," "his," or "hers")
- `RB`: Adverb, very (for example, "silently")
- `RBR`: Adverb, comparative (for example, "better")
- `RBS`: Adverb, superlative (for example, "best")
- `RP`: Participle (for example, "give up")
- `TO`: For example, to go "to" the store
- `UH`: Interjection (for example, "errrrrrrm")
- `VB`: Verb, base form (for example, "take")

- VBD: Verb, past tense (for example, "took")

- VBG: Verb, gerund/present participle (for example, "taking")

- VBN: Verb, past participle (for example, "taken")

- VBP: Verb, sing. present, non-3D (for example, "take")

- VBZ: Verb, third-person sing. present (for example, "takes")

- WDT: Wh-determiner (for example, "which")

- WP: Wh-pronoun (for example, "who" or "what")

- WP$: Possessive wh-pronoun (for example, "whose")

- WRB: Wh-abverb where, (for example, "when")

To apply part-of-speech tagging in Optimus, use the following:

```
print(df.cols.pos("text"))
```

This will give us a tuple for every word in every row, as shown here:

```
text
(object)
[('transformers', 'NNS'), ('american', 'JJ'), ('japanese',
'JJ'), ('media', 'NNS'), ('franchise', 'NN'), ('produced',
'VBN'), ('american', 'JJ'), ('toy', 'NN'), ('company', 'NN'),
('hasbro', 'VBZ')
...
```

The following summarizes the results:

- 'transformers' and 'media' were detected as 'NNS', meaning "noun plural."

- 'american' and 'japanese' were detected as 'JJ', meaning "adjective."

- 'produced' was detected as 'VBN', meaning "past participle."

- 'toy' and 'company' were detected as 'NN', meaning "noun, singular."

- 'hasbro' was detected as 'VBZ', meaning "third-person singular."

Part-of-speech tagging in itself may not be the solution to any particular NLP problem. However, it is something that is done as a prerequisite to simplify a lot of different problems such as text-to-speech conversion and word-sense disambiguation.

Applying the transformation

Now that we have a good understanding of all the functions, let's apply them together. Remember that you can chain the functions, as shown in the following example:

```
(df
    .cols.strip_html()
    .cols.remove_urls()
    .cols.remove_special_chars()
    .cols.remove_stopwords()
    .cols.lemmatize_verbs()
    .cols.num_to_words()
    .cols.word_tokenizer())
```

This will transform our data to this:

```
text
(object)
['transformer', 'american', 'japanese', 'medium', 'franchise',
'produced', 'american', 'toy', 'company', 'hasbro', 'japanese',
'toy', 'company', 'takara', 'tomy', 'follows', 'battle',
'sentient', 'living', 'autonomous', 'robot', 'often',
'autobots', 'decepticons', 'transform', 'form', 'vehicle',
'animal', 'franchise', 'encompasses', 'toy', 'animation',
'comic', 'book', 'video', 'game', 'film', 'franchise', 'began',
'one', 'thousand', ',', 'nine', 'hundred', 'and', 'eighty-
four', 'transformer',
...
```

Now that we have normalized and removed all the possible noise from our string, it is time to transform our text to features that can be used as input for creating our machine learning models.

Feature extraction from text

When using text in machine learning, we need to convert text to a list of features a machine learning algorithm can understand. This means that we need to convert text to numbers. To accomplish this, there are two approaches that can be used with Optimus:

- Bag of words
- TF-IDF

Let's see how you can use these methods in Optimus.

Bag of words

In the bag of words approach, we take all the words and then count the number of occurrences of each word.

After counting the number of occurrences of each word, because a corpus can have millions of words, it can be useful to select the most frequent word in the text, as shown in the following figure:

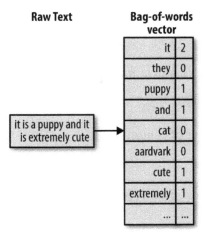

Figure 9.2 – Bag of words example

To apply bag of words in Optimus, you can use the following code:

```
_df = df.cols.bag_of_words("text")
```

This returns a big dataframe with all the strings as column names and the word count in every row. Because it can be difficult to explore all the data, let's print the first 10 values using the following code:

```
_df.cols.names()[:10]
```

This will return the following:

```
['1983',
 '1984',
 '1991',
 '2001',
 '2005',
 '2007',
 '2010s',
 '2018',
 '2019',
 '20th']
```

To get the whole count, we can use the following code:

```
len(_df.cols.names())
```

That will return the total number of columns:

```
139
```

To print a sample of the whole dataframe, you could use _df:

1983	1984	1991	2001	2005	2007	2010s	2018	2019	20th	aligned	alternate	american	animals
1	2	3	4	5	6	7 (int64)	8	9	10	11 (int64)	12 (int64)	13 (int64)	14 (int64)
(int64)	(int64)	(int64)	(int64)	(int64)	(int64)		(int64)	(int64)	(int64)				
not nullable	not nullable	not nullable	not nullable	not nullable	not nullable	not nullable	not nullable	not nullable	not nullable	not nullable	not nullable	not nullable	not nullable
1	1	1	1	1	2	1	1	1	1	1	1	2	1

Figure 9.3 – The displayed result of the dataframe in a notebook

As you can see, the words are sorted in alphanumeric order in this new dataframe, with the number of words per value.

Let's see how to expand the version of bag of words with n-grams.

Using bigrams or trigrams over unigrams (words)

An extended version of the bag of words approach is the bag of n-grams approach. An n-gram is simply any sequence of *n* tokens (words). It can be 1-grams, 2-grams, 3-grams, and so on.

Let's see an example for every case using this text:

```
from nltk import ngrams
sentence = 'Transformers is the most amazing TV series.'
```

For 1-grams, it will output tuples with one value:

```
print(list(ngrams(sentence.split(), 1)))
[('Transformers',), ('is',), ('the',), ('most',), ('amazing',),
('TV',), ('series.',)]
```

For 2-grams, it will output tuples with two values, where the last value of the tuple is the first in the continuous tuple:

```
print(list(ngrams(sentence.split(), 2)))
[('Transformers', 'is'), ('is', 'the'), ('the', 'most'),
('most', 'amazing'), ('amazing', 'TV'), ('TV', 'series.')]
```

Let's see what happens with 3-grams:

```
print(list(ngrams(sentence.split(), 3)))
[('Transformers', 'is', 'the'), ('is', 'the', 'most'), ('the',
'most', 'amazing'), ('most', 'amazing', 'TV'), ('amazing',
'TV', 'series.')
```

It is important to point out that bag of n-grams can be more informative than a simple bag of words because it captures more information about the context of each word. For example, (`'most'`, `'amazing'`, `'TV'`) can give you more information than just (`'amazing'`,). However, this comes at a cost, as bag of n-grams can produce a much larger and sparser feature set than bag of words (filtering methods help to minimize this). Typically, 3-grams is about as high as we want to go, as using n-grams beyond that rarely increases performance because of sparsity.

To control the kind of n-gram you want to produce, just use the `ngram_range` parameter as shown in the following code:

```
df.cols.bag_of_words("text", ngram_range= 2)
```

This will result in a dataframe with columns representing 2-grams, as follows:

Viewing 1 of 1 rows / 184 columns

1 partition(s)

1983 tonka 1 (int64) not nullable	1984 transformers 2 (int64) not nullable	1991 subsequently 3 (int64) not nullable	2001 idw 4 (int64) not nullable	2005 third 5 (int64) not nullable	2007 distinct 6 (int64) not nullable	2007 liveaction 7 (int64) not nullable	2010s attempt 8 (int64) not nullable	2018 transformers 9 (int64) not nullable	2019 incarnations 10 (int64) not nullable	20th century 11 (int64) not nullable
1	1	1	1	1	1	1	1	1	1	1

Viewing 1 of 1 rows / 184 columns

1 partition(s) <class 'optimus.engines.pandas.dataframe.PandasDataFrame'>

Figure 9.4 – HTML representation of the result of the method called

As we can see, using 2-grams allows us to count phrases instead of just words. This can provide a complex insight into the data.

Now let's see how TF-IDF works.

TF-IDF

Another way to weigh the frequency of words in a document is by using TF-IDF.

TF-IDF is a numerical statistic that is intended to reflect how important a word is to a document in a collection or corpus. TF-IDF returns values between 0 and 1 for every word. Words that appear in many documents have a value closer to zero and words that appear in fewer documents have values closer to 1.

For this example, let's load a file with a couple of rows. Each row represents a document in TF-IDF:

```
df = op.load.file("text-e.csv")
```

Let's take a look at the data. The following command will print a dataset resume:

```
print(df)
```

The dataset resume is shown here:

```
text(object)
```

```
"<a>http://google.com</a> <span>that'll Transformers is an
American and Japanese media franchise produced by American toy
company Hasbro and Japanese toy company Takara Tomy. It follows
the battles of sentient, living autonomous robots, often the
Autobots and the Decepticons, who can transform into other
```

forms, such as vehicles and animals. The franchise encompasses
toys, animation, comic books, video games and films. The
franchise began in 1984 with the Transformers toy line,
comprising transforming mecha toys from ...

Applying TF-IDF in Optimus is as simple as invoking the `tf_idf` method:

```
df.cols.tf_idf("text").to_dict()
```

This will return a dictionary with a lot of values:

```
{'1980s': [0.0, 0.07302266125566928],
 '1983': [0.028561620660523312, 0.0],
 '1984': [0.02032182833693315, 0.05195622490690069],
 '1991': [0.028561620660523312, 0.0],
 '1993': [0.0, 0.07302266125566928],
 '2001': [0.028561620660523312, 0.0],
 '2005': [0.028561620660523312, 0.0],
 '2007': [0.057123241321046625, 0.0],
 '2010s': [0.028561620660523312, 0.0],
 '2018': [0.028561620660523312, 0.0],
 '2019': [0.028561620660523312, 0.0],
 '20th': [0.028561620660523312, 0.0],
 'able': [0.0, 0.07302266125566928],
 'action': [0.057123241321046625, 0.0],
 'after': [0.057123241321046625, 0.0],
 'afterward': [0.0, 0.07302266125566928],
 'again': [0.057123241321046625, 0.0],
 'aligned': [0.028561620660523312, 0.0],
 'also': [0.0, 0.07302266125566928],
 'alternate': [0.028561620660523312, 0.0],
 ...
```

In this case, we're using `5to_dict` to get a better insight into the data. For big dataframes
like this, remember that you can filter out some specific columns. Let's say that we want to
filter the `toys` and `1984` columns:

```
df.cols.tf_idf("text")[["1984","toys"]].print()
```

This will print the following:

```
        1984         toys
     (float64)    (float64)
   -----------   -----------
     0.0203218    0.0406437
     0.0519562    0.0519562
```

As we can see, the output only includes the two columns passed to the brackets.

By using TF-IDF, we can get useful information from our data. For example, we could use it to find out how popular a topic is in the comment section of a website in comparison to others.

Summary

In this chapter, we covered all the functions available in Optimus to easily clean and prepare your text data so you can start your NLP journey, from simple operations, such as removing stopwords and URLs, to more advanced ones, such as stemming and lemmatization.

We learned how to tokenize and tag the text in our datasets to efficiently capture information from them.

After that, we explored a couple of methods to get features from text. We saw how to use bag of words and TF-IDF to convert text to numbers that can be used as input to machine learning algorithms.

In the next chapter, we will cover what we consider to be Optimus' most advanced features, such as implementing your engine, creating custom data transformation functions, and even plotting functionality.

10
Hacking Optimus

So far, we have covered almost everything Optimus can do for you, from data loading and data preparation to machine learning and natural language processing.

In this chapter, you will deep dive into Optimus and learn how to expand what Optimus can do for you. We will be covering the following topics:

- Adding a new engine
- Bumblebee
- Joining the community
- The future
- Limitations

Technical requirements

Optimus can work with multiple backend technologies to process data, including GPUs. For GPUs, Optimus uses **RAPIDS**, which needs an NVIDIA card. For more information about the requirements, please go to the *GPU configuration* section of *Chapter 1, Hi Optimus!*.

You can find the code for this chapter at https://github.com/ PacktPublishing/Data-Processing-with-Optimus.

Installing Git

If you don't have Git installed, please go to the link that matches your operating system. This will help you download the latest Optimus code to your machine:

- Windows: `https://git-scm.com/download/win`
- Linux and Ubuntu/Debian, from the CLI: `sudo apt install git-all`
- Mac: `git -version` and follow the instructions provided

Now, let's learn how to add a new engine to Optimus.

Adding a new engine

The ultimate thing you can do with Optimus is expand its functionality to make it do whatever you want. In this section, we will explore how Optimus is structured so that can know where to go to expand or add specific functionality. To make this a complete exercise and understand where every piece goes, we'll add a completely new engine, such as pandas, Dask, cuDF, or Dask-cuDF. Here, we'll be adding Vaex as the engine of choice.

Cloning the repository from GitHub

To start, let's clone a clean Optimus repository from GitHub. Be aware that the file path and Optimus's internal structure may change a little in upcoming releases:

1. First, let's clone the Optimus repository. From the CLI, enter `git clone https://github.com/ironmussa/Optimus.git`.

    The easiest way to test the new code in Optimus is by using a couple of handy comments from a Jupyter Notebook.

    In the first cell, add the following:

    ```
    %load_ext autoreload
    %autoreload 2
    ```

    This will reload the Optimus code every time it changes automatically.

2. After that, insert the path to the folder where you cloned Optimus:

    ```
    import sys
    sys.path.insert(0, "../Optimus")
    ```

Now that we have the code in place, let's explore how it is organized.

How the project is organized

Most of the engines supported by Optimus have more functions in common than not. Here, we can create some base classes that function as the core implementation. This will form most of Optimus's functionality.

In the /optimus/engines/base path, we defined an abstract function and the base implementation for loading and saving data, column and row functions, masks, and dataframe creation.

Then, in every folder in /optimus/engines/, while following the engine's naming convention, implement specific functionality for every engine:

```
/pandas. Pandas implementation
/dask. Dask Implementation
/cudf. Cudf Implementation
/dask_cudf. Dask cudf implementation
/spark. Spark Implementation
/vaex. Vaex implementation.
/ibis. Ibis implementation.
```

By deep diving inside each of these engine folders, you can find the specific folder structure you need. The idea here is to implement the methods that are not compatible with the base implementation.

In every engine folder, you can find the following:

```
/io. Loading, Saving and Database handling
/ml. Machine Learning Classes
/create.py. Dataframe creation.
/columns.py. Columns Functions.
/rows.py. Rows Functions.
/functions.py. Base functions that support rows a columns
function. String, Math and Trigonometric functions.
/engine.py. Handle client creation
/mask.py. Mask functions
```

Now that we know the general structure, let's explore how it works.

The entry point

When you start Optimus, you see that one of the main functions is def optimus, which can be found in the /Optimus/optimus/optimus.py file. To add a new engine, we need to do the following:

1. Add the new engine you want to add to the Engine class, as shown in the following code block:

```
class Engine(Enum):
    PANDAS = "pandas"
    CUDF = "cudf"
    DASK = "dask"
    DASK_CUDF = "dask_cudf"
    SPARK = "spark"
    VAEX = "vaex"
    IBIS = "ibis"
```

Here, we added a constant called vaex.

2. After that, in the Optimus class, add the following code:

```
elif engine == Engine.VAEX.value:
    from optimus.engines.vaex.engine import VaexEngine
    op = VaexEngine(*args, **kwargs)
```

In Optimus, all the implementation engines are in Optimus/optimus/engines/. In the preceding code, we imported the VaexEngine implementation, but we haven't implemented anything yet.

3. To implement this, we will create the engine.py file and add the following code:

```
from optimus.engines.base.engine import BaseEngine
from optimus.engines.vaex.create import Create
from optimus.engines.vaex.io.load import Load
from optimus.optimus import Engine
from optimus.version import __version__
```

```
import vaex


class VaexEngine(BaseEngine):
    __version__ = __version__

    def __init__(self, verbose=False):
        self.verbose(verbose)
        self.client = vaex

    @property
    def create(self):
        return Create(self)

    @property
    def load(self):
        return Load(self)

    @property
    def engine(self):
        return Engine.VAEX.value
```

All the engines we create must inherit from the `BaseEngine` class. This class will enforce the fact that you need to implement the `Create`, `Load`, and `Engine` classes.

4. Now, we need to implement the following in these functions:

    a) In the `Create` class, the dataframe functions are implemented to create dataframes using Python dictionaries and pandas. This is implemented in `Optimus/optimus/engines/vaex/create.py`.

    b) In the `Load` class, all the functions are implemented to load data files that are in CSV, JSON, and XLS format. This is implemented in `Optimus/optimus/engines/vaex/io/load.py`.

    c) The `Engine` class will return the string describing the engine. This is the string you created in the `Engine` class.

As an example, let's implement a really simple `load csv` function:

```
def csv(path):
    dfd = vaex.read_csv(path)
    return VaexDataFrame(dfd)
    df.meta = Meta.set(df.meta, value={"file_name": path,
"name": ntpath.basename(path)})
```

As you can see, we need to return a `VaexDataFrame`. Let's learn how to implement `VaexDataFrame`. For this, we must create a file called `Optimus/optimus/engines/vaex/io/dataframe.py`.

5.   The first thing we must do is implement `VaexDataFrame`, which will inherit `BaseDataFrameClass` and enforce the implementation of 15+ functions such as `.cols`, `.rows`, and `.save` to enforce a consistent API for all the engines supported by Optimus:

```
from optimus.engines.base.basedataframe import
BaseDataFrame
class VaexDataFrame(BaseDataFrame):
    def __init__(self, data):
        super().__init__(self, data)
```

6.   To finish, let's test the `load` function:

```
df = op.load.csv("foo.csv")
```

7.   Now, we can print our dataframe:

```
print(df)
name           function
(string)       (string)
----------     ------------------------
Optimus        leader
Bumblebee      espionage
eject          ELECTRONIC SURVEILLANCE
```

Now that we've created the Vaex Dataframe, let's learn how to add a function that will allow us to operate over string and numeric data.

Base class functions

One of the core Optimus classes is the `Functions` class. It abstracts and implements some core functionality that you can rely on when implementing another engine. Let's see how it works.

String transformations

In the case of the Vaex implementation, and because it has some similarities with the pandas dataframe, you could get some of the `cols` functions for free.

Let's try the lower operations so that we can convert all the strings into lowercase.

As the Vaex operation is the same as in all the other dataframes, using the `.str` accessor, we can rely on the function that's already been implemented in `BaseClass`:

```
print(df.cols.upper("name", output_cols="a"))
```

Also, you will have the `print` implementation for free. As usual, it will print your dataset:

name number	new_column	function	phone_
(string)	(string)	(string)	(string)
---------- ----------	------------	------------------------	-----------
OPTIMUS	OPTIMUS	leader	123-456-7890
BUMBLEBEE	BUMBLEBEE	espionage	123-456-7890
EJECT cybertron.com	EJECT	ELECTRONIC SURVEILLANCE	optimus@

Let's explore how to implement a function that is not compatible with the ones we've already implemented. In Vaex 4.0, `slice` does not have a `step` parameter, so it is going to produce an error. Say you try to use the following code:

```
df.cols.slice("name",sbtart=1, stop=3)
```

This will throw an error, similar to the following:

```
TypeError: str_slice() takes from 1 to 3 positional arguments
but 4 were given
```

To fix this, we will use the base `Functions` class as a skeleton to create our Vaex functions. Now, let's create the `VaexFunctions` class in the `Optimus/optimus/engines/vaex/functions.py` file:

```
from optimus.engines.base.functions import Functions
class VaexFunctions(Functions):
    def slice(self, series, start, stop, step):
        # Step is not handle by Vaex for version 4.0
        return series.str.slice(start, stop)
```

If you take a close look, you will see that the `step` parameter is present in the function. We will work like this to preserve and not break the compatibility it has with other engines.

To finish, add the newly created `Functions` class to the Vaex DataFrame. Just go to `/Optimus/optimus/engines/vaex/dataframe.py` and add the following method:

```
@property
def functions(self):
    from ptimus.engines.vaex.functions import VaexFunctions
    return VaexFunctions()
```

Now, you can execute the following command:

```
print(df.cols.slice("name",start=1, stop=3))
```

You will receive the following sliced column:

name	function	phone_number
(string)	(string)	(string)
----------	------------------------	---------------------
pt	leader	123-456-7890
um	espionage	123-456-7890
je	ELECTRONIC SURVEILLANCE	optimus@cybertron.com

Now that we have learned how to operate over string data, let's learn how to add a function to operate over numeric data.

Numeric transformations

In the previous example, we learned how to map a Vaex function so that it can be used in Optimus. Now, we will learn how to implement some trigonometric functions.

Depending on the engine you want to implement, some functionality may be part of the engine core's functionality, while others may depend on external libraries such as NumPy.

This is the case with Vaex.

An important thing about how Optimus handles some operation is that it ensures that the data type (string, numeric, or date) you are working on will work with the function that you want to apply. In this example, we'll learn how to implement a `sine` function that only works with numeric data.

As you already know, the Vaex functions are implemented in `Optimus/optimus/engines/vaex/functions.py`. Since we've already implemented a skeleton, let's add the `sine` function. Be sure to add the `numpy` library at the top of the file:

```
import numpy as np
....
def sin(self, series):
    return np.sin(self.to_float(series))
```

Now, apply `sin` to the `num` columns and select it:

```
print(df.cols.sin("num")["num"])
```

We will get the following output:

```
        num
    (float64)
    -----------
     0.808496
     0.14112
    -0.756802
```

Now that we are done with numeric transformations, let's cover applying functions next.

Applying functions

When applying a function to a dataframe, Optimus provides you with three options:

- **Vectorized**: This method is fast, and operations are applied to multiple pieces of data at the same time.

- **Partitioned**: This is applied to partitioned data that's used in Dask.

- **Mapped**: This method is slow. The operations area is applied to every element.

Optimus provides these different modes because not all functions can be vectorized. Next, we will learn how to implement a function using every case.

To handle these three operations, Optimus implements the `apply` function.

In contrast to pandas, where the `apply` function is not vectorized, Optimus has the `mode` parameter, which you can us to select how to apply a function.

In the case of `slice` and `sin`, which we have already implemented, they are vectorized under the hood. And surprise – the actual Optimus logic can handle this without any changes needing to be made to the code. In the case of mapped operations, we need to implement a little logic. Let's take a look:

1. Optimus implements the _map class in every engine to handle these cases. For Vaex, go to the `/Optimus/optimus/engines/vaex/columns.py` file and create the _map function:

```
from optimus.engines.base.dataframe.columns import
DataFrameBaseColumns


class Cols(DataFrameBaseColumns):
    def __init__(self, df):
        super(DataFrameBaseColumns, self).__init__(df)


    def _map(self, df, input_col, output_col, func,
*args):
        return df.apply(func, arguments=(df[input_col],
*args,), vectorize=False)
```

2. Now, let's create a function with a very simple operation that will add 10 to our num columns:

```
def func(value):
    return value + 10
print(df.cols.apply("num",func, mode="map")["num"])
```

This will return the following output:

```
        num
   (float64)
 -----------
      12.2
      13
      14
```

Now that you have learned how to add functions that modify data in your dataframe, let's take a step back and learn how to load data.

I/O operations

Think about that shiny new data format that you want to use but hasn't been implemented in Optimus yet. This data format is probably already implemented in some of the engines. So, let's learn how to add this functionality to Optimus.

Loading data

To add new loading functionality, you need to address two important points:

- Return a Vaex dataframe
- Metadata

When loading data in any format, you should use the engine to load the data, create the Optimus `VaexDataFrame` object, and return it. Let's take a look:

1. First, you must create the `Load` class in `/Optimus/optimus/engines/vaex/io/load.py`. As an example, let's create a function that will load an HDF5 file:

```python
class Load(BaseLoad):
    @staticmethod
    def hdf5(path, columns=None, *args, **kwargs):
        path = unquote_path(path)
        dfd = vaex.open(path)
        df = VaexDataFrame(dfd)
        df.meta= Meta.set(df.meta, value={"file_name":
path, "name": ntpath.basename(path)})
        return df
```

As you can see, we are passing a normal Vaex DataFrame to the `VaexDataFrame` class to create an Optimus DataFrame. With this, we are giving the original Vaex dataframe all the methods available in Optimus, such as `cols`, `rows`, and `mask`.

One important point is that by convention, Optimus preserves some extra data in the `meta` property. Here, we will save the filename and the path in the `df.meta` property using the `Meta` class.

2.   That's all that we need to do to create a data loader in Optimus. To load and only show the column name data, you can use the following command:

```
print(op.load.hdf5("foo.hdf5").cols.select("function"))
```

You will get the following output:

```
function
(string)
-----------------------
leader
espionage
```

Remember that you can always access the Vaex original data form by using the following command:

```
df.data
```
```
To get the data dataframe.  #  name        function
phone_number                  num
   0   't'      ' leader'                    '123-456-7890'
2.2
   1   'm'      ' espionage'                 '123-456-7890'
3
   2   'e'      ' ELECTRONIC SURVEILLANCE'   'optimus@
cybertron.com'     4
```

Now, let's learn how to save the data that we worked with.

Saving data

Let's suppose that we want to save our data in another HDF5 file.

The file for implementing this class is in the same folder that the Load class it in, but in the save.py file; that is, /Optimus/optimus/engines/vaex/io/save.py:

1. A simple function we can implement is as follows:

    ```
    class Save():
        def hdf5(self, path, conn, *args, **kwargs):
            df = self.root.data
            df.export_hdf5(path, *args, **kwargs)
    ```

2. Here, we are simply getting the Vaex dataframe using .data from the root param. root is a special property that's used to access the highest level of a dataframe.

3. After that, we must simply call the Vaex method to save the file.

4. To finish, let's add our new create class to the VaexDateFrame constructor.

5. In /Optimus/optimus/engines/vaex/dataframe.py, add the following to the top of the file:

    ```
    from optimus.engines.vaex.io.save import Save
    ```

6. Then, add the save method to the VaexDataFrame class:

    ```
    @property
        def save(self):
            return Save(self)
    ```

7. Now, we can use the following command to save our dataframe in the new_foo.hdf5 file:

    ```
    df.save.hdf5("new_foo.hdf5")
    ```

Now, we'll move on to plotting the data.

Plots

As we saw in *Chapter 5, Data Visualization and Profiling*, Optimus can handle four types of plots associated with specific visualizations:

- Histogram and frequency using bar plots
- Scatter plots
- Box plots
- Correlations

You can create the `plot` function in `Optimus/optimus/plots/plots.py`. Here, you can add any methods you want to the `Plot` class; for example:

```
class Plot:
    def my_new_shiny_chart(a,b,c):
        # process your data using optimus functions
        # plot your data using matplotlib
```

The plotting function is built over the `Optimus` function and uses the same code for all the engines. When converting the data into a pandas dataframe, you have a lot of options for plotting. The problem here is that you may not be able to handle them locally because of resource constraints. Using Optimus functions for the already implemented engines ensures that you can plot your chart without any problems arising.

After processing your data, you can use matplotlib (already an Optimus requirement) to plot the data as required.

Since the `plot` method is implemented in the `BaseDataFrame` class, you don't need to add it to the implementation of `VaexDataFrame`.

Profiler data types

The profiler data types are data types that are inferred by Optimus to help you understand your data. As we mentioned in *Chapter 1, Hi Optimus!*, Optimus brings some profiler data types out the box, such as URL, email, and credit cards, besides the convectional int, floats, and dates.

In Optimus, you can create new profiler data types. The entry point is the `infer_profier_dtypes` function in `Optimus/optimus/engines/base/columns`.

This calls the `infer_dtypes` function, which is in charge of inferring every value data type, for example, `infer_dtypes` in `Optimus/optimus/engines/base/functions`. This function is to be added step by step:

```
if is_list(value):
dtype = ProfilerDataTypes.ARRAY.value
elif is_null(value):
dtype = ProfilerDataTypes.NULL.value
elif is_bool(value):
dtype = ProfilerDataTypes.BOOLEAN.value
elif is_credit_card_number(value):
dtype = ProfilerDataTypes.CREDIT_CARD_NUMBER.value
```

```
elif is_zip_code(value):
dtype = ProfilerDataTypes.ZIP_CODE.value
...
```

As you can see, every sentence is trying to infer a data type. The order in which the places are put can be tricky, so you may need to modify the way the tree is constructed.

The best way to implement your own profiler data types, as well as your own parser, is by inheriting and reimplementing these functions.

Let's say we want to implement a phone number data type. For this, we could use a regular expression to check if a string is a phone number in a specific format:

- The following is an example of this:

```
def is_phone_number(value):
return re.match("^[+]*[(]{0,1}[0-9]{1,4}[)]{0,1}[-\s\./0-
9]*$", value)
```

- In our `if` tree, we need to add this function like so:

```
...
elif is_phone_number(value):
dtype = "phone_number"
```

- The `dtype` string is going to be our identifier so that Optimus can count and infer the `column` data type. Ensure that the string you use doesn't match any that are used by Optimus. To check the string with Optimus, you can check the `ProfilerDataTypes` class in the `Optimus/optimus/helpers/contants` file: `df = op.load.csv("foo.csv")`:

name	function	phone_number
(object)	(object)	(object)
----------	-----------------------	--------------------
--		
Optimus	leader	123-456-7890
Bumblebee	espionage	123-456-7890
eject com	ELECTRONIC SURVEILLANCE	optimus@cybertron.

- Now, to test this, we will use the following code:

```
print(df.cols.infer_dtypes())
```

This will print every data type that's been detected in every cell in the dataframe, like so:

name	function	phone_number
(object)	(object)	(object)
----------	----------	---------------
string	string	phone_number
string	string	phone_number
string	string	email

Inside Optimus, this is used to infer the data type in a column that counts all the data types that were detected in the column's cells.

Expanding Optimus can give you a lot of flexibility. Let's explore a special use case, that is, how to add functionality to Optimus so that you can build a user interface.

Bumblebee

Bumblebee is an open source, low-code web app that aims to make big data preparation easy. It builds on top of Optimus so that you have all the flexibility the library provides.

Bumblebee sends automatically generated Optimus code to a Python kernel gateway to operate our datasets and configuration settings. All of this is done over a secure connection.

For example, when we ask Bumblebee to load a file, it automatically uploads the file to a place Optimus can find it (since it may not be able to load the file from your local storage) and loads it using `op.load.file`.

Bumblebee has a broad range of available operations, and almost every Optimus function is mapped as a user-friendly interface.

In the web app, we can take advantage of its profiling functionality to give the user insight into the loaded data. This also includes loading the actual values of the dataset into a table in real time:

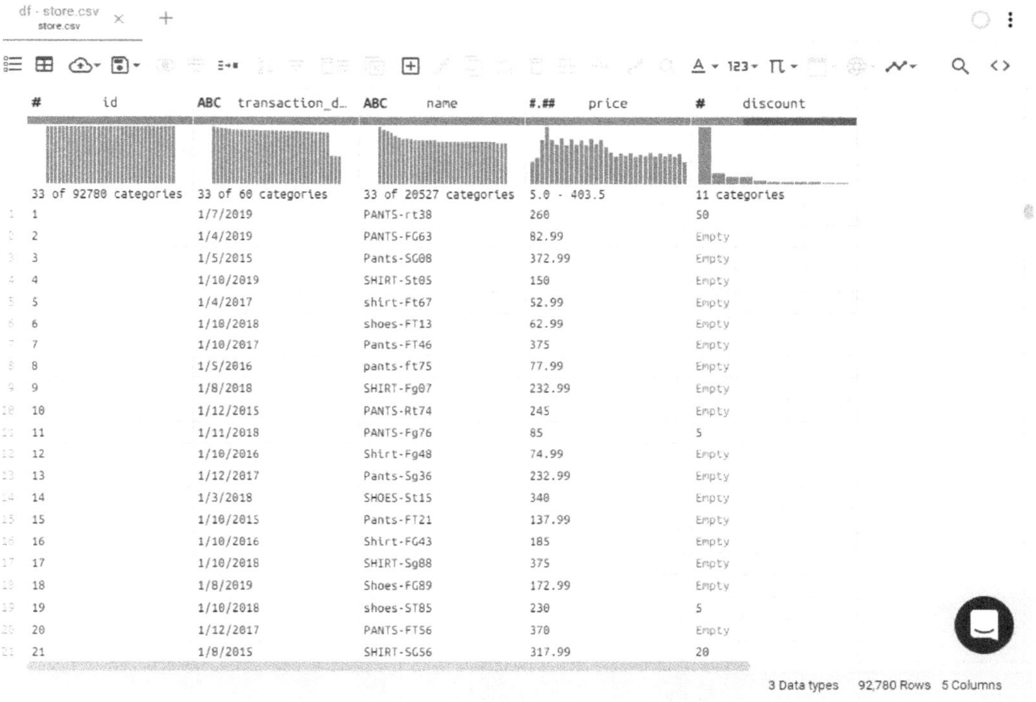

Figure 10.1 – Bumblebee default view with a loaded dataframe

As we can see, once the data is loaded, the user will instantly get useful information, such as the quality of the data, the distribution of any frequent values, and the first rows of the dataset.

Before confirming some of the transformations that are available in Bumblebee, we can look at a real-time preview of the data. This is useful for testing:

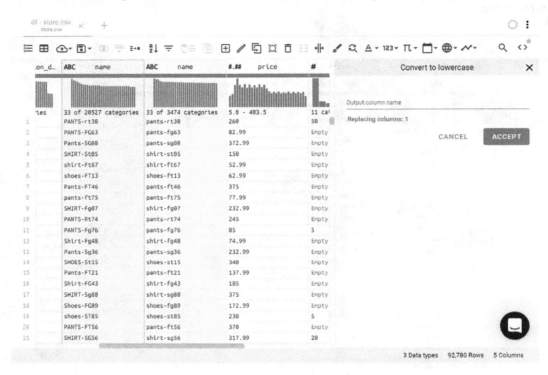

Figure 10.2 – Preview of a lowercase operation

And of course, we can save the dataset locally, into remote file storage, or simply download it via our browser.

The web app also allows you to export all the operations that were made in the Optimus code, in case we want to migrate it to our advanced workflow:

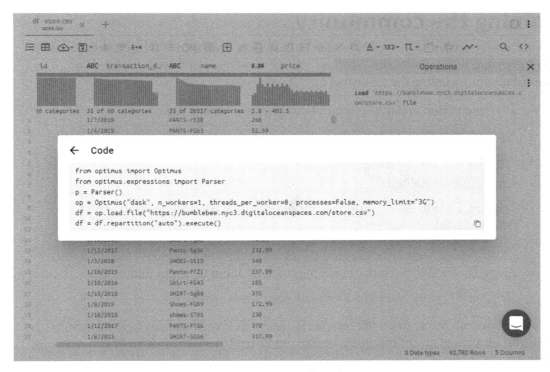

Figure 10.3 – Exporting code to Optimus

Bumblebee has other features, such as user management, workspace management, engine configuration management, column visualization filters, and column searching.

Bumblebee and Optimus are open source projects, which means you can contribute to them by performing pull requests, commits, and more.

Bumblebee was created by the same team that built Optimus. We designed Optimus so that it can be easily extended so that you can build things such as a full-featured frontend interface.

We are eager to see what people can build with Optimus, so don't hesitate in joining the community if you have an idea or want to build something amazing with this project.

Joining the community

As we have stated throughout this book, Optimus is an open source project. Contributions go far beyond pull requests and commits. We are very happy to receive any kinds of contributions, including the following:

- Documentation updates, enhancements, designs, or bug fixes
- Spelling or grammar fixes
- `README.md` corrections or redesigns
- Unit or functional tests
- Triaging GitHub issues, especially for determining whether an issue persists or is reproducible
- Searching for `#optimusdata` on Twitter and helping someone else who needs help
- Blogging, speaking about, or creating tutorials about Optimus and its many features
- Helping others on Slack at `slack.hi-optimus.com`

We are always on Slack, so do not hesitate to reach out if you need any help or want to share an idea about any of our projects. You can also open a GitHub issue if you need any help with Bumblebee.

The future

The main goal of Optimus is to help you find the easiest way to wrangle data, regardless of it data format/source or the infrastructure available to you. Taking that into account, we are going to continue building a strong foundation of methods and almost all the functions you will ever need to wrangle your data.

Bumblebee was the second obvious step since it provides an easy-to-use UI to load, explore, and wrangle data that's been built over Optimus. Our plan with Bumblebee is to continue to preserve our vision of the overall Optimus project. We will continue to put open source and easy to use tools in user's hands so that they can get insight into their data as quickly and easily as possible.

The third step was the Optimus API. With the Optimus API, we aimed to give every user access to the power of Optimus, regardless of the programming language being used. From JavaScript in the browser to Node.js to C++, you can access the power of any of the Optimus engines from anywhere. We are sure that users will find creative and amazing ways to use it.

We are very excited about the future of Optimus and Bumblebee. To continue building a product that is useful for you, your feedback is vital. So, once again, we invite you to keep telling us about yourself and how you use – or plan to use – Optimus.

Limitations

We are working hard to create a unified API from the most popular dataframe libraries. However, all the technologies are in different development stages. Many issues have been flying under the radar, but here, we want to highlight some of the most important ones.

Right now, the main limitations are as follows:

- Creating a UDF for string processing in cuDF and Dask-cuDF since they are not supported yet: `https://github.com/rapidsai/cudf/issues/7301`.
- cuDF, Dask-cuDF, and Vaex database connections are handled using Dask, which needs to load data as pandas dataframes and then convert them into the appropriate format based on a certain engine.
- Regex cuDF support is limited. For example, it still can handle lowercase and uppercase characters at the same time: `https://github.com/rapidsai/cudf/issues/5217`.

Summary

In this chapter, we did a deep dive into how Optimus was built and helped you learn how to expand what Optimus can do, from creating a new engine to loading and saving data and processing string and numeric data.

This can give you a lot of flexibility. You also learned how to build a full-fledged frontend interface in which you can use Optimus as the backend to process data on any of the engines that are available.

In the next chapter, we will talk about the Optimus server and how you can use it to process data using an easy-to-use API.

If you find any other inconsistencies in the API, please let us know.

11
Optimus as a Web Service

There's growing demand for web applications and services in general. Nowadays, almost anything you can think of can be done by using just a computer and an internet browser. So, why should data transformation lag behind?

In this chapter, we will learn about how Optimus can be applied in a way that can help our data wrangling process to be even more seamless. We can also use what we learn to apply an easy way of data wrangling in our own web tools.

The topics we will be covering in this chapter are as follows:

- Introducing Blurr
- Setting up the environment
- Making requests
- Optimus' features

Technical requirements

Optimus can work with multiple backend technologies to process data, including GPUs. For GPUs, Optimus uses RAPIDS, which requires an NVIDIA card. For more information about the requirements, please go to the *GPU configuration* section in *Chapter 1.*

Blurr has very specific requirements. To learn more, visit `https://github.com/hi-primus/optimus`.

> **Important**
>
> The Blurr project is still in development stages, but the beta version will be available to the readers by the time the book is published. You can refer to the GitHub repo (`https://github.com/hi-primus/optimus`) for more updates.

You can find all the code in this chapter at `https://github.com/PacktPublishing/Data-Processing-with-Optimus`.

Introducing Blurr

When creating a new project, tools play an important part in the development process. Because of this, we created Blurr, a package that provides us with a friendly API to make requests in a data wrangling context.

Blurr provides a handful of different options for how we can use it to develop or sketch a new project. It is even useful for quickly wrangling a file without digging into it using Python with Optimus; just a browser and an address are needed to request some transformations.

When using this tool, communication in a Python environment will be handled automatically. All that is needed is an available Jupyter Kernel Gateway address (we'll see more about this later):

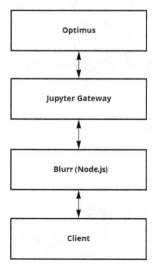

Figure 11.1 – How Blurr and Optimus communicate

By making Blurr available in an endpoint, you'll be able to wrangle your data from any kind of environment using any language that supports HTTP requests, opening a range of possibilities.

Let's learn what projects can be made using Blurr:

- A data wrangling web application
- A file format transforming service

Also, Blurr can be used on already existing projects to add features such as these:

- Creating database dumps
- Generating reports from collected data
- Detecting discrepancies and outliers in your data

By configuring with just a few lines of code, you can also activate features that will allow you to quickly request transformations by using JSON. You can take advantage of this for debugging.

Now that we know what we can do, let's learn how to set up and use this package.

Setting up the environment

Let's see how we can set up Blurr in a JavaScript environment.

Pre-requisites for Blurr

Before installing Blurr, we must know what **Node.js** is and how its package manager works. Just having Node.js and **NPM** installed should be enough. To install these packages, please go to the official documentation here: `https://nodejs.org/en/docs/guides/getting-started-guide/`.

The Node.js installation will include NPM, which will already take us halfway. We're now able to initialize a new Node.js project using `npm init` on the command line. Now let's learn how to install the package.

Also, we're going to need **Jupyter Kernel Gateway** running in WebSocket mode. To learn how to set up and initialize an instance, read the documentation available here: `https://jupyter-kernel-gateway.readthedocs.io/`.

Installing the package

To install Blurr in an already set-up Node.js project, we can use the following command in our Terminal or Command Prompt:

```
npm install hi-blurr
```

For yarn, use this:

```
yarn add hi-blurr
```

Once it's done, we'll have the package available for us to use in our Node.js project, but first we'll need it imported in our code.

Importing the package

To use Blurr on our code, we can import it using require or import:

```
let { Blurr } = require("blurr")
```

The preceding code will make the Blurr class available to us. Let's see how we can import it using ES6 imports in the following code:

```
import { Blurr } from "blurr"
```

This will also provide us with the Blurr class. Let's learn how to initialize an instance with which we'll be able to start processing our datasets.

Creating a Blurr session

To create a Blurr session, we can run the following code:

```
let session = new Blurr({
    kernelAddress: "localhost:8888",
    downloadsFolder: "./download",
    downloadsAddress: "http://localhost:3000/download/"
    engineConfiguration: {
        engine: "dask",
        memory_limit: "2 GB"
    }
});
```

In the preceding code, we created a new `Blurr` instance and passed it to a variable named `session`; in this instantiation, we're passing the address of a Jupyter Kernel Gateway instance running in WebSocket mode that should be capable of running Optimus.

The other parameter passed in the previous example is `engineConfiguration`, in which you can select what engine you are willing to use. In this case, we're passing `"dask"`, which is the default engine, so in this case, we can omit `engineConfiguration`. Refer to the following code:

```
let session = new Blurr({
    kernelAddress: "localhost:8888"
});
```

The previous code will behave the same as the earlier example.

We also defined `downloadsFolder` and `downloadsAddress` to use a previously configured endpoint to handle our downloads. We'll discover more about that later in this chapter.

We can use the created instance to start making requests, but first, let's see what we can do if we want to use two engines in the same session.

Multiple engines in one session

By using the session instance, we can create multiple engines, as shown in the following code:

```
blurr = session.engine({
    engine: "dask",
    memory_limit: "2 GB"
})
blurrPandas = session.engine({
    engine: "pandas"
})
```

In the preceding code, we're instantiating two engines in the same session named `blurr` and `blurrPandas`. The variable session in the previous example has the same.

If we don't want to specify certain parameters, we can simply use the name of the engine as the only argument, as shown in the following code:

```
blurr = session.engine("dask")
blurrPandas = session.engine("pandas")
```

The last two examples will behave the same except for the one named `blurr`, which will have some extra configuration in the first case.

Quickest setup

By default, you can pass all the methods available in an engine instance to a session instance, so to quickly create an engine using **dask**, you can use the following code:

```
let blurr = new Blurr({ kernelAddress: 'localhost:8888' })
```

In the preceding example, we're simply creating a session, but in this case, we'll use it as an engine since it will create one by default once the first request is made.

Now we have learned how to create sessions and engines, we can start making requests.

Making requests

To make a request, there's a method called `request` available in every session or engine instance using Blurr. It accepts a JavaScript object as an argument to support compatibility with JSON and make it web-friendly.

The following is an example:

```
let response = await blurr.request({
    operation: "createDataframe",
    dict: {
        id: [0, 1, 2, 3],
        value: [1.5, 2.25, 11, 12.5]
    },
    saveTo: "df"
});
```

We used `await` to wait for the value of the promise. The `.then()` method can be used as well. Either way, depending on the request, you may get the same response – either a complete response or a status of what was changed. If we check the content of `response` after running the previous code, we'll get the following result:

```
{
    status: "ok",
    updated: "df",
```

```
    code: "df = op.create.dataframe({\"id\": [0, 1, 2, 3],
\"value\": [1.5, 2.25, 11, 12.5]}) "
}
```

In this case, we're getting a status telling us whether the request was successful, what variables were updated, and the code that was executed.

If we don't pass a `saveTo argument`, Blurr will assign a variable name automatically. This only applies to requests that don't transform other dataframes but instead create them; otherwise, it'll overwrite the input dataframe. We'll look at this later in this chapter. Refer to the following code:

```
await blurr.request({
    operation: "createDataframe",
    dict: { test: [0, 1, 2, 3] }
});
```

We'll get the following output:

```
{
    status: "ok",
    updated: "df0",
    code: "df0 = op.create.dataframe({\"test\": [0, 1, 2, 3]}) "
}
```

As is evident, Blurr assigned the name df0 to the dataframe created. Now let's learn how to load a file instead of creating a dataframe from scratch.

Loading a dataframe

As you noticed in the previous example, we used an argument named `dict` that is specific to the `createDataframe` operation. If instead you wish to load a file, you can use an operation called `loadFile` and pass the address to `path`. Let's see how in the following example:

```
let response = await blurr.request({
    operation: "loadFile",
    path: "Chapter 11/example.csv",
    saveTo: "df"
});
```

In the preceding example, we're asking Blurr (and therefore, Optimus) to load a file stored in `"Chapter 11/example.csv"`. In this case, it will be loaded from the local storage of the remote kernel used to run Optimus on Python:

```
{
    status: "ok",
    updated: "df",
    code: "df = op.load.file(path=\"Chapter 11/example.csv\")"
}
```

The preceding response tells us what was updated, the status of the request, and the code that was executed in the kernel.

We can also use connections to, for example, load files from S3 filesystems. We'll see more about that later in this chapter.

Let's see how we can save a dataframe.

Saving a dataframe

As with loading, you can save a dataframe into a file.

In this case, we'll be saving to the local Jupyter instance. To do this, you can simply use the following request:

```
await blurr.request({
    operation: "saveFile",
    dfName: "df",
    path: "Chapter 11/example-saved.csv"
});
```

The preceding code will save the same dataframe into `"Chapter 11/example-saved.csv"`, which will give us a similar response:

```
{
    status: "ok",
    code: "df.save.csv(path=\"Chapter 11/example-saved.csv\")"
}
```

In this case, there are no dataframes updated, but we still get the status and the code executed. As we can see, we didn't specify what file type we were going to save. In this case, Blurr extracted this from the name of the file, but if we want to set the file type, we can pass it as an argument as shown in the following code:

```
await blurr.request({
    operation: "saveFile",
    dfName: "df",
    path: "Chapter 11/example-saved.xls",
    type: "excel"
});
```

Or we can do so by calling a specific method, such as "saveParquet", as shown here:

```
await blurr.request({
    operation: "saveParquet",
    dfName: "df",
    path: "Chapter 11/example-saved.parquet"
});
```

In both preceding examples, we would have a similar response. But what if we want a link to download the dataset instead? Let's see a special operation for this case in the following code:

```
await blurr.request({
    operation: "downloadFile",
    dfName: "df",
    path: "Chapter 11/example-saved.csv"
});
```

This request will work only if the Jupyter Kernel Gateway instance and the Node.js process are running on the same machine and your Node.js process has an endpoint configured for downloads. Refer to the following:

```
{
    status: "ok",
    url: "localhost:3000/public/Chapter 11/example-saved.csv",
    code: "df.save.csv(path=\"Chapter 11/example-saved.csv\")"
}
```

As we can see, the code in the response is not the real code executed, so it doesn't expose the real location of the downloads folder. This is for security reasons.

To load and save from a remote filesystem or a database, we'll have to learn how to handle connections. This will be addressed at the end of this section. Before that, let's learn how we can explore the dataframe we loaded.

Getting information from the dataset

The following is an example of a request with an output that won't be just a status response:

```
await blurr.request({
    operation: "countRows",
    dfName: "df"
});
```

In the preceding example, we're requesting the number of rows of this dataset. For that, we passed `"countRows"` as the name of the operation, but since we're coming from Optimus, we can simply enter the name of the method with its accessor, as shown in the following code:

```
await blurr.request({
    operation: "rows.count",
    dfName: "df"
});
```

This can be applied to methods available on Optimus instances. The content of both responses will be as follows:

```
{
    status: "ok",
    content: 12,
    code: "df.rows.count()"
}
```

The attribute of the response named `content` will contain the output of the operation requested. Let's see another example with a different response:

```
await blurr.request({
    operation: "columns",
```

```
        dfName: "df"
});
```

This will give us the following output:

```
{
    status: "ok",
    content: ["id", "name", "value"],
    code: "df.cols.names()"
}
```

In this case, we're requesting a list of every column. Now, content contains an array of strings.

But what if we want to transform the columns in our dataframe instead? Let's see some examples.

Transforming a dataset

Transforming is no different from what we already know, except for the dfName parameter. That shall contain the name of the dataset we already defined, as shown in the following code block:

```
await blurr.request({
    operation: "upper",
    columns: "name"
    dfName: "df"
});
```

Also, we can pass what we received from a previous response:

```
let dataframe = await blurr.request({
    operation: "loadFile",
    path: "Chapter 11/example.csv"
});
response = await blurr.request({
    operation: "upper",
    columns: "name"
    dfName: dataframe
});
```

This will extract the name from the `updated` attribute of the response.

The responses to the last two requests will be as follows:

```
{
    status: "ok",
    updated: "df",
    code: "df = df.cols.upper(columns=\"name\")"
}
```

By default, it'll save the transformed dataframe into the same input dataframe. If you want to save it on another, just use `saveTo` as shown here:

```
await blurr.request({
    operation: "lower",
    columns: "name"
    dfName: "df",
    saveTo: "df_lower",
});
```

This will give us the following output:

```
{
    status: "ok",
    updated: "df",
    code: "df_lower = df.cols.lower(columns=\"name\")"
}
```

In the preceding example, we're creating a new variable for the transformed dataframe with the name `df_lower`. Also, we're passing an argument directly into the request. Let's see some alternatives for how we can pass some arguments into our requests.

Passing arguments

Arguments can be passed into requests by using keywords, such as `columns: "name"`, but what if we want to pass unnamed arguments?

For that, we can use `args`, as shown in the following code block:

```
await blurr.request({
    operation: "min",
```

```
    dfName: "df",
    args: ["value"]
});
```

This will give us the following output:

```
{
    status: "ok",
    content: 0.5,
    code: "df.cols.min(\"value\")"
}
```

As you can see, we passed an argument without using keywords. This is just for compatibility reasons. Usually, you may just pass the argument as an attribute, in this case, `columns`, as follows:

```
await blurr.request({
    operation: "min",
    dfName: "df",
    columns: "value"
});
```

This will behave the same as the example before. You can also pass arguments in a different attribute, named `kwargs`:

```
await blurr.request({
    operation: "min",
    dfName: "df",
    kwargs: { columns: "value" }
});
```

As demonstrated, you have plenty of alternatives to achieve the same request. We recommend not using `args` or `kwargs`, but depending on your need, you may want to use one of them.

But what if you want to check the content of your dataset rather than just getting an insight into it? Let's see how you can do that.

Getting the content of the dataset

Blurr has a method that allows us to get this content in a Python dictionary. In the response, the content will be shown as a JavaScript object with one array per column. Let's look at the following code:

```
await blurr.request({
    operation: "display",
    dfName: "df",
    n: 4
});
```

This will give us the following results:

```
{
    status: "ok",
    content: {
        id: [1, 2, 3, 4],
        name: ["John", "Alice", "Bob", "Carol"],
        value: [5.25, 3.5, 0.5, 1.75]
    },
    status: "ok",
    code: "df.to_dict(n=4)"
}
```

This is useful to get a quick view of the content of the dataset.

Let's see what we can do to make slightly more complex requests by passing multiple operations.

Multiple operations in one request

By passing an array, you can use the same request for multiple operations:

```
await blurr.request([{
    operation: "createDataframe",
    dict: { A: [0, 1] },
    saveTo: "df1"
}, {
    operation: "display",
```

```
    dfName: "df1"
}]);
```

We'll get the following result:

```
{
    status: "ok",
    content: {
        A: [0, 1]
    },
    status: "ok",
    code: "df = op.create.dataframe({\"A\": [0, 1]}); df.to_
dict()"
}
```

In this response, we got the content of the dataframe we just created. This represents the content of the last operation passed to the request.

We have learned how to handle dataframes by using requests. Let's now learn how to handle connections (see *Chapter 2*) and clusters made by using string clustering methods (see *Chapter 6*).

Using other types of data

When using Optimus, we don't just create variables to save dataframes; we also create variables for connections to remote file storage systems and for the results of a string clustering process. Let's see how Blurr handles this.

Connections

To create connections, you can simply use any of the operations available with connect<Type>, where <Type> is the type of connection to be created. This will behave the same as loadFile or createDataframe, creating a variable and returning a status response. Let's see an example of creating a connection to an S3 bucket using connectS3:

```
await blurr.request({
    operation: "connectS3",
    endpoint_url: "http://region.example.com",
    bucket: "example"
    saveTo: "conn"
}
```

This will give us the following response:

```
{
    status: "ok",
    updated: "conn",
    code:_"conn_=_op.connect.s3(endpoint_url=\
    "http://region.example.com\",_bucket=\"example\")"
}
```

Since we entered the data to create the connection, we won't need to check its content, but that's not the case for string clusters, which are calculated.

String clusters

String clusters (data that is internally stored in dictionaries) allow us to get feedback on the clusters and to use the methods available in the `Clusters` class.

Let's use `stringClustering` on a dataframe:

```
await blurr.request([{
    operation: "loadFile",
    path: "Chapter 11/names.csv",
    saveTo: "df2"
}, {
    operation: "stringClustering",
    dfName: "df2",
    columns: "name"
}]);
```

This will give us a response that includes some content that will be useful for replacing the values:

```
{
    status: "ok",
    updated: ["df2", "clusters"],
    code: "df2 = op.load.file(path=\"Chapter 11/names.csv\")\n
clusters = df2.string_clustering(\"name\")\n clusters",
    content: [
```

```
        {suggestion: "Alice", suggestions: ["Alice", "alice"],
suggestions_size: 2, total_count: 4},
        {suggestion: "Bob", suggestions: ["Bob", "bob"],
suggestions_size: 2, total_count: 2},
    ]
}
```

As we can see, we passed two operations in one request. The second one will calculate and return the clusters and save them in a variable named `clusters`. This name is assigned automatically, since we did not pass any value to `saveTo` in the second operation.

To replace a suggestion from a cluster, you can use `setSuggestion`, as shown in the following example:

```
await blurr.request({
    operation: "setSuggestion",
    suggestion: "Alice",
    value: "Alice Doe"
});
```

We'll get the following response:

```
{
    status: "ok",
    updated: "clusters",
    code: "clusters.set_suggestion("Alice", "Alice Doe")",
}
```

In the previous request, we adjusted the value of the cluster so that it could be applied to our dataset by passing the name of the variable to the `clusters` parameter.

We have learned how to load, save, and transform our data, and how to get certain information from it, but the examples shown are not the only things you can do using Blurr. Next, we'll see what features are available on Blurr.

As we can see, most of the names are the same, and others are just simply the name of the method in camel case rather than in snake case (for example, `patternCounts` instead of `pattern_counts`).

To get a complete list of the operations and features from Optimus available on Blurr, you can view the documentation at `https://hi-bumblebee.gitbook.io/blurr/`.

Summary

In this chapter, we learned about Blurr, a library that provides us with an alternative to Python environments that can instead be used in a typical web environment, opening up a range of possibilities.

We learned how to set the tool up, how to use it, and how it compares with Optimus, including some of the operations seen in previous chapters of this book, so we can use them in this alternative workspace.

We looked at things such as loading, saving, profiling, and transforming. More advanced features such as connections and clustering were also illustrated, giving us a clearer view of what can be done using Blurr.

`Packt.com`

Subscribe to our online digital library for full access to over 7,000 books and videos, as well as industry leading tools to help you plan your personal development and advance your career. For more information, please visit our website.

Why subscribe?

- Spend less time learning and more time coding with practical eBooks and Videos from over 4,000 industry professionals

- Improve your learning with Skill Plans built especially for you

- Get a free eBook or video every month

- Fully searchable for easy access to vital information

- Copy and paste, print, and bookmark content

Did you know that Packt offers eBook versions of every book published, with PDF and ePub files available? You can upgrade to the eBook version at `packt.com` and as a print book customer, you are entitled to a discount on the eBook copy. Get in touch with us at `customercare@packtpub.com` for more details.

At `www.packt.com`, you can also read a collection of free technical articles, sign up for a range of free newsletters, and receive exclusive discounts and offers on Packt books and eBooks.

Other Books You May Enjoy

If you enjoyed this book, you may be interested in these other books by Packt:

Python Data Cleaning Cookbook

Michael Walker

ISBN: 978-1-80056-566-1

- Find out how to read and analyze data from a variety of sources
- Produce summaries of the attributes of data frames, columns, and rows
- Filter data and select columns of interest that satisfy given criteria
- Address messy data issues, including working with dates and missing values
- Improve your productivity in Python pandas by using method chaining
- Use visualizations to gain additional insights and identify potential data issues

Python Data Analysis – Third Edition

Avinash Navlani, Armando Fandango, Ivan Idris

ISBN: 978-1-78995-524-8

- Explore data science and its various process models
- Perform data manipulation using NumPy and pandas for aggregating, cleaning, and handling missing values
- Create interactive visualizations using Matplotlib, Seaborn, and Bokeh
- Retrieve, process, and store data in a wide range of formats
- Understand data preprocessing and feature engineering using pandas and scikit-learn
- Perform time series analysis and signal processing using sunspot cycle data
- Analyze textual data and image data to perform advanced analysis

Packt is searching for authors like you

If you're interested in becoming an author for Packt, please visit `authors.packtpub.com` and apply today. We have worked with thousands of developers and tech professionals, just like you, to help them share their insight with the global tech community. You can make a general application, apply for a specific hot topic that we are recruiting an author for, or submit your own idea.

Share Your Thoughts

Now you've finished *Data Processing with Optimus*, we'd love to hear your thoughts! Scan the QR code below to go straight to the Amazon review page for this book and share your feedback or leave a review on the site that you purchased it from.

`https://packt.link/r/1-801-07956-0`

Your review is important to us and the tech community and will help us make sure we're delivering excellent quality content.

Index

www.ingramcontent.com/pod-product-compliance
Lightning Source LLC
Chambersburg PA
CBHW060516060326
40690CB00017B/3299